NONLINEAR WAVES

NONLINEAR WAVES

Edited by SIDNEY LEIBOVICH
and A. RICHARD SEEBASS

CORNELL UNIVERSITY PRESS
ITHACA AND LONDON

International Standard Book Number 0-8014-0766-4
Library of Congress Catalog Card Number 72-12285

Contents

   Generalized Solutions

IV. Examples of Dissipative and Dispersive Systems
    Leading to the Burgers and the Korteweg-deVries
    Equations   103
    Sidney Leibovich and *A. Richard Seebass*

    1. Perturbation Techniques   104
    2. Dissipative Gas Dynamics and the Burgers Equation   112
    3. Shallow-Water Waves and the Korteweg-deVries   125
       Equation

V. Dispersive Waves and Variational Principles   139
   Gerald B. Whitham

    1. Dispersive Waves   140
    2. Fourier Synthesis and Asymptotic Behavior   145
    3. Simple Derivation of Group Velocity Concepts for   149
       Linear Problems
    4. Extensions and Examples   152
    5. Variational Principles   155
    6. Adiabatic Invariants, Wave Action, and Energy   162
    7. Nonlinear Group Velocity—Stability of Periodic   164
       Waves
    8. Formal Perturbation Theory   166

VI. Conservation of Wave Action   170
    Wallace D. Hayes

    1. Discrete Dynamic Systems   171
    2. Action Conservation for a Family of Solutions   173
    3. Steady Oscillating System   173
    4. Action Conservation for a Slowly Varying System   176
    5. Continuous Media   178
    6. Waves   180
    7. Comparison of Action and Energy Conservation   183
    8. Wave Propagation Theory   184

VII. Wave Interactions   186
     Owen M. Phillips

    1. The Resonance Conditions   188
    2. The Interaction Equations   192
    3. Interactions among Surface Gravity Waves   198

Contents

Preface

The existence of "waves" as a recognizable form of disturbance is one of the most familiar features of the physical world; their description is one of the unifying threads connecting diverse areas of physics. Major gaps still exist, however, in the mathematical theory of waves. A well-developed body of information is available only for those waves described (usually approximately) by linear equations or by quasilinear hyperbolic differential equations.

Recent developments in the theory of nonlinear wave propagation have occurred rapidly and have been far-reaching. In an effort to acquaint graduate students and faculty members in engineering and the physical sciences with these developments, the editors arranged, in 1969, a special series of seminars on nonlinear wave propagation under the auspices of the Center for Applied Mathematics and the College of Engineering. The seminars, consisting of introductory talks by Cornell faculty members and lectures by distinguished scientists and mathematicians from other universities, form the basis of this book. The text is derived from the manuscripts submitted by the speakers, supplemented by other invited contributions.

There are reasons to believe that a published version of the seminars will be of considerable value. No coherent account of this subject matter exists in the literature, although certainly more than enough is now known to make a collection of these ideas worthwhile. This volume brings together the fundamental ideas that have shaped the recent work and are likely to determine fruitful directions for future research.

Since the purpose of the seminar series was to provide a genuine educational experience, we attempted to make the lectures appropriate for beginning graduate students in engineering. A tutorial and well-motivated development of the subject was sought and, it is felt, achieved. The same approach was requested of the authors of this book.

No widely accepted classification of nonlinear waves exists. Still, a rough separation can be made into nondispersive motions, which frequently can be described by quasilinear hyperbolic partial differential equations, and dispersive motions. Most of the recent advances concern dispersive motion.

Much of the newer work may be divided, somewhat arbitrarily, into two categories. In the first class of problems, ad hoc singular perturbation techniques or averaging methods are employed to construct a uniformly valid solution; the aim is to correct a linearized solution in order to account for the cumulative effect of weak nonlinearities, or to modify a nonlinear solution in order to account for small departures from periodicity, or to allow for some other modulation. In the second group of problems, the exact behavior of solutions of model equations is studied. These model equations usually arise as approximations to more complicated systems.

Most of the concepts that arise are introduced and illustrated in the introductory chapter by Wallace D. Hayes. The two categories of this newer work are covered in more detail later in the book, after careful treatments of the linear dispersive theory by Stephen A. Thau and of quasilinear hyperbolic systems by Constantine M. Dafermos. It is probably worth emphasizing that hyperbolic systems continue to play a major role in nonlinear dispersive systems, since the propagation of quantities such as wavenumber and frequency is controlled by hyperbolic systems. A treatment of weakly nonlinear waves first appears in Chapter IV, where the two-timing (or multiple-scale) perturbation procedures are briefly reviewed and applied to derive the Burgers and the Korteweg-deVries equations. Solutions and properties of these model equations, some of which are needed later in the book, are recorded in this chapter.

Gerald B. Whitham, in Chapter V, develops the method of averaged variational principles that he pioneered. This line of attack has had a powerful impact on work in nonlinear dispersive wave behavior and forms a major portion of the second category of the work mentioned. In Chapter VI the concepts and results introduced by Whitham are presented, from a slightly different point of view, by Hayes, who emphasizes the fundamental role of "wave action." Owen M. Phillips introduces, in Chapter VII, the concept of resonant wave interactions in weakly nonlinear systems and outlines how these interactions can be computed. Chapter VIII, by Robert M. Miura, is divided into two parts. In the first part the transformation introduced by Gardner, Greene, Kruskal, and Miura is explained; this procedure replaces the nonlinear Korteweg-deVries equation with a linear integral equation and produces explicit exact solutions. In the second part a two-time WKB type of perturbation is applied to the Korteweg-deVries equation in an attempt to describe behavior in the limit of vanishing dispersion.

In many circumstances the asymptotic multiple-time structure of a problem is straightforward. This is not always the case, however, and in Chapter IX, Richard E. Meyer reviews a process of finding solutions when a double limit is involved. Ideas developed by Kaplun for boundary-layer problems are used, and one might speculate that in the future the several singular perturbation methods now known will share a common mathematical base. The study of stratified fluids has provided great stimulus for the development of theories of wave motion, and a rich variety of possibilities has emerged. In Chapter X, Chia-Shun Yih develops aspects of infinitesimal and finite-amplitude waves in stratified fluids: in addition to the linearized results, an important class of steady nonlinear wave solutions is discussed. The final chapter, by Peter D. Lax, is devoted to more rigorous investigations of nonlinear model equations. The main emphasis here is upon understanding the mathematical basis for the success of the transformation of Gardner et al. for the Korteweg-deVries equation so that similar results may be found for other equations. It seems likely that the ideas described in this chapter point the way for significant lines of research.

Chapter V, some figures in Chapter VII, and a substantial portion of Chapter XI have been reprinted with the kind permission of the Mathematical Association of America, the Royal Society of London, and the University of Tokyo Press.

We are grateful to the authors for their contributions. Although some of our requests may have seemed tedious, the authors remained willing and helpful throughout, and this eased our task. We owe thanks to our wives for their patient tolerance of too many evenings and weekends without our company, to Barbara Stevens for editorial assistance, to Barbara Benedict and Mary Gehres for typing assistance, and to Carol Wilser for the line drawings. Finally, we are indebted to the Center for Applied Mathematics and the College of Engineering at Cornell and to the National Science Foundation for the funds that made the seminar series, and therefore this book, possible.

S. L.
A. R. S.

Ithaca, New York

NONLINEAR WAVES

CHAPTER I

Introduction to Wave Propagation

Wallace D. Hayes

1. Oscillatory Motions

Most wave motions are essentially oscillations, oscillations that propagate in space. To describe them we require as independent variables the time t and at least one space coordinate x. Since the equations governing wave propagation involve at least two independent variables, the equations generally involve partial derivatives.

Many of the properties of propagating waves are also properties of simple oscillating systems governed by ordinary differential equations, with time as independent variable. While propagation in space is not one of these properties, of course, a number of the properties of waves may be illustrated by a study of a simple linear oscillator.

1a. The Linear Oscillator

The simplest oscillator comprises a mass m that can move in one direction. It is constrained by a linear spring of spring constant K. The displacement of the mass from an equilibrium position is y (Figure I.1). The force on the mass is then $-Ky$ and is equal to the time derivative of the momentum $d(my_t)/dt = my_{tt}$. The subscripts here and later denote derivatives with respect to the independent variables.

The basic frequency ω_0 of the oscillator is defined by

$$\omega_0{}^2 = \frac{K}{m}. \tag{1}$$

This chapter is based upon a series of lectures given by the author for the University of Washington's program "Modern Engineering for Managers" in the summer of 1967. Permission to use this material is gratefully acknowledged.

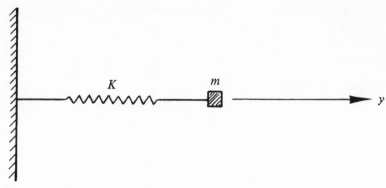

I.1. Linear oscillator.

The basic equation governing the oscillator is then

$$y_{tt} + \omega_0^2 y = 0. \tag{2}$$

The general solution to equation (2) is

$$y = y_1 \cos \omega_0 t + y_2 \sin \omega_0 t = y_0 \cos(\omega_0 t - \alpha), \tag{3}$$

where y_1, y_2, $y_0 = (y_1^2 + y_2^2)^{1/2}$ and α are constants. We introduce a variable θ which we term the *phase*, given here by

$$\theta = \alpha - \omega_0 t. \tag{4}$$

The solution is sinusoidal in the phase, and the phase is linear in the time.

In linear problems, with the solutions sinusoidal in a phase variable, it is convenient to introduce a complex solution, with the solution of interest being represented by the real part. Thus the solution (3) may be written

$$\begin{aligned} y &= \mathrm{Re}\,\{(y_1 + iy_2)e^{-i\omega_0 t}\} \\ &= \mathrm{Re}\,\{y_0 e^{i\theta}\}. \end{aligned} \tag{5}$$

It is customary to write the solution as a complex one, without the operation of taking the real part. However, the real part *must* be taken before calculating energy of any kind.

The kinetic energy of the oscillator is given by

$$T = \tfrac{1}{2} m y_t^2 = \tfrac{1}{2} m \omega_0^2 y_0^2 \sin^2 \theta = \tfrac{1}{2} K y_0^2 \sin^2 \theta.$$

The potential energy stored in the spring may be expressed

$$V = \tfrac{1}{2} K y^2 = \tfrac{1}{2} K y_0^2 \cos^2 \theta.$$

The potential energy is always defined to be zero if the system is undisturbed. The total energy is the sum of these two, or

$$E = T + V = \tfrac{1}{2}K y_0^2.$$

This quantity is a constant, and energy, therefore, is conserved.

We denote by $\bar{T}$ and $\bar{V}$ the averages of T and V over a complete cycle, as θ decreases by 2π ($360°$). The average of $\sin^2 \theta$ is $\tfrac{1}{2}$, and the average of $\cos^2 \theta$ is also $\tfrac{1}{2}$. We can write, then,

$$\bar{T} = \bar{V} = \tfrac{1}{4}K y_0^2 = \tfrac{1}{2}E.$$

This result says that, *on the average*, half the energy of the oscillator is kinetic and half is potential. This property is referred to as *equipartition*. If the Lagrangian L is defined as $L = T - V$, this property may be expressed as $\bar{L} = 0$.

If the oscillator is damped, a force other than the spring force acts in the direction opposite to the velocity y_t. The simplest case is that of linear damping, with the damping force proportional to the velocity. We define the damping parameter β by the condition that the damping force be $-2m\beta y_t$. The basic equation then reads

$$y_{tt} + 2\beta y_t + \omega_0^2 y = 0. \tag{6}$$

The general solution to this equation takes the form

$$y = y_0 e^{-\beta t} \cos (\omega t - \alpha),$$

where

$$\omega = (\omega_0^2 - \beta^2)^{1/2},$$

provided that $\beta < \omega_0$. We do not consider the "overdamped" case with $\beta \geq \omega_0$.

In this case the total energy is not constant, but decreases. Because of this property the system is termed *dissipative*. Roughly speaking, the energy decays as $e^{-2\beta t}$; the actual law is somewhat more complicated. If $(\beta/\omega_0) \ll 1$, the system is termed weakly dissipative. In this case the frequency ω is very close to ω_0, the average energies $\bar{T}$ and $\bar{V}$ can still be reasonably defined, and equipartition may be said to still hold.

1b. The Pendulum and Nonlinearity

A simple example of a nonlinear oscillator is that of the pendulum. We consider a mass m suspended by a massless string of length l and with motion in a plane (Figure I.2). The moment of inertia about the pivot or suspension point is ml^2. The restoring torque when the mass has been displaced by an angle ϕ is $mgl \sin \phi$. The resulting basic equation is

$$\phi_{tt} + \omega_0^2 \sin \phi = 0, \tag{7}$$

where

$$\omega_0^2 = \frac{g}{l} \tag{8}$$

again gives the basic frequency.

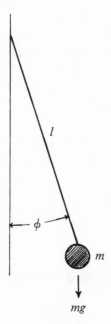

I.2. Pendulum.

If the maximum value of $\phi = \phi_0$ is small, we can approximate $\sin \phi$ by ϕ. Equation (7) is then the same as (2) for the linear oscillator. In fact, by letting $y = l \sin \phi \approx l\phi$, we can interpret the lateral motion of the mass as though it were attached to a spring of spring constant $K = m\omega_0^2 = mg/l$.

When we may no longer make the simplifying assumption that $\sin \phi \doteq \phi$, then the basic equation (7) is nonlinear (and nondissipative). The solution may be expressed in terms of elliptic integrals; the details would take us too far away from the subject of waves. Energy is still conserved in this case, but equipartition between kinetic and potential energies no longer holds. If the energy is less than $2mgl$, the motion is still oscillatory, with the frequency no longer constant but dependent upon the energy. The motion is no longer sinusoidal. (If $E > 2mgl$, the motion is rotary.)

The energy is $E = mgl(1 - \cos \phi_0)$, and one approach is to carry out an expansion in ϕ_0. This must be done in such a way as to take the change of frequency into account (e.g., by the Poincaré method; see Cole, 1968). One result can be quoted for the frequency,

$$\frac{\omega}{\omega_0} = 1 - \frac{\phi_0^2}{16} + O(\phi_0^4).$$

We consider next the same pendulum, with the linearizing approximation $\sin \phi = \phi$, but now permitted to swing in two directions. With the equivalent spring constant $K = mg/l$ and with ω_0^2 given by (1) or (8), the equations governing the motion are

$$y_{tt} + \omega_0^2 y = 0,$$
$$z_{tt} + \omega_0^2 z = 0.$$

The first of these equations is the same as (2), as is the second with y replaced by z.

We now have two dependent variables, which represent components of a vector in a two-dimensional euclidean space. The two equations are *uncoupled*, by which we mean that z does not appear in the first and y does not appear in the second. This implies that the two equations may be solved independently. Each represents a linear oscillator. The system is termed *degenerate*, meaning that the frequencies of the two linear oscillators are the same, so that two independent modes of motion with the same frequency are possible.

In this case the phenomenon of *polarization* appears. A solution with $z = 0$ is polarized in the y direction, one with $y = 0$ is polarized in the z direction, and one with $z = y$ or with $z = -y$ is polarized at $45°$. A solution

$$y = y_0 \cos (\omega_0 t - \alpha),$$
$$z = y_0 \sin (\omega_0 t - \alpha)$$

is circularly polarized and follows a circular path on the (y, z) plane in a counterclockwise direction. The same solution with z changed in sign is also circularly polarized but moves in the clockwise direction.

The same results can be obtained with a mass as in Figure I.1 but with motion permitted in a plane and with a second spring of the same spring constant K mounted at right angles to the first.

If the equations are nonlinear, they generally become coupled through the nonlinearity. This is the case for the spherical pendulum without the linearizing assumptions. As with the planar pendulum, the frequency is no longer independent of amplitude.

1c. Summary of Properties of Simple Oscillators

We present here a list of characteristic properties of simple oscillators that are also possessed by propagating waves. For waves, we shall have a phase θ, which is a function of position in space as well as of time. It will vary linearly with x in a one-dimensional case so that (4) becomes

$$\theta = kx - \omega t + \alpha. \tag{9}$$

The quantity k is termed the wavenumber, and is $2\pi\lambda^{-1}$, where λ is the wavelength of a sinusoidal wave.

The characteristic properties of linear nondissipative systems, except where otherwise indicated, are:

(i) The dependent variables vary sinusoidally with respect to a phase θ, which is linear in the time.

(ii) The frequency $\omega = -\partial\theta/\partial t$ is given by a unique relation characteristic of the system (also of the wavenumber in the case of propagating waves) and *not* of the particular solution.

(iii) Total energy is conserved.

(iv) On the average, energy is equally divided between kinetic and potential energy.

(v) If two dependent uncoupled variables represent components of a vector in a plane and the system is degenerate with respect to these variables, the solution may possess a particular type of polarization.

(vi) If the system is weakly dissipative, the amplitudes and the total energy decrease slowly but exponentially, while the other properties above are essentially unchanged.

(vii) If the system is strongly dissipative (but not overdamped), the properties above fail or are substantially altered.

(viii) If the system is nonlinear but nondissipative, total energy is conserved. The other properties of linear nondissipative systems fail.

2. Linear Nondissipative Propagating Waves

We consider next the general nature of a system of waves propagating in a one-dimensional space. The independent variables are now t and x. There are one or more dependent variables describing the wave system; here we consider only one and label it y. The effects of nonlinearity or of strong dissipation are primarily to destroy the simple regular properties in which we are interested. The effect of weak dissipation is primarily a factor $e^{-\beta t}$ in the amplitudes or $e^{-2\beta t}$ in the energies; we will consider it later. Accordingly, we limit ourselves to the case of linear nondissipative wave propagation. Actual physical systems exhibiting wave propagation are discussed in Section 4. Here we consider certain features shared by all linear nondissipative propagating wave systems.

2a. Phase and Phase Velocity

The general form of the solution that we expect for the dependent variable $y(x, t)$ is the same as that of (5),

$$y = y_0 \, \text{Re}\{e^{i\theta}\} = y_0 \cos \theta, \tag{10}$$

where θ is the phase. The phase is linear in the independent variables, or

$$\theta = kx - \omega t + \alpha. \tag{11}$$

We have here repeated equation (9). The phase will be constant for an observer who moves with the velocity

$$\frac{dx}{dt} = \frac{\omega}{k} = c. \tag{12}$$

This *phase velocity* measures the velocity of an individual crest ($\theta = 2\pi m$), trough ($\theta = \pi + 2\pi m$), or node ($\theta = \frac{1}{2}\pi + \pi m$) of y. Note that

$$\frac{d\theta}{dt} = \frac{\partial \theta}{\partial t} + \frac{dx}{dt} \frac{\partial \theta}{\partial x} = 0$$

if dx/dt is given by (12). Here, of course, m is any integer.

2b. Dispersion

The property of propagating waves analogous to property (ii) of linear oscillators is that a functional relation $f(\omega, k) = 0$ connects the frequency and the wavenumber. This is usually expressed in the form

$$\omega = W(k), \tag{13}$$

with the frequency expressed as a function of the wavenumber k. The phase velocity c is then given as a function of wavenumber:

$$c(k) = \frac{\omega}{k} = k^{-1} W(k).$$

Of particular importance is the special case in which $c(k)$ is constant and the phase velocity is independent of the wavenumber (or of the frequency). In this case, which is termed *nondispersive*, the function W is simply

$$W(k) = kc. \tag{14}$$

If the phase velocity is not constant, the wave propagation is termed *dispersive*. A relation of the form (13) is generally termed a *dispersion relation* or, in certain contexts, an eikonal equation.

In nondispersive wave propagation, the property analogous to the sinusoidal property (i) of linear oscillators becomes essentially unimportant. Assume we know that we have only waves that move with phase velocity c and that the distribution of $y(x, 0)$ at $t = 0$ is known. If this initial distribution is *not* sinusoidal, we may represent it through a Fourier integral as a superposition of sinusoidal distributions. Since the system is supposed linear, each of these component sinusoidal distributions develops following (10) and (11), with frequency given by (14). At a later time t, the solution may be obtained by reconstituting the Fourier integral.

In the nondispersive case this process recovers the original distribution translated by a distance ct. In other words, the solution may be expressed as

$$y(x, t) = y(x - ct, 0).$$

Another interpretation is that the sinusoidal solution $y_0 \cos \theta$ may be replaced by

$$y = f(\theta) = f(kx - \omega t + \alpha) = y(x - ct, 0)$$

in terms of an arbitrary function of the phase defined using some reference values of k and ω. The quantity $x - ct$ may be considered a choice of phase corresponding to the choice of reference $k = 1$, $\omega = c$. The phase is now unimportant except as an independent variable in terms of which the unvarying signal is expressed.

If the wave propagation is dispersive, this type of argument fails completely. Each Fourier component must be considered separately in this case.

2c. Group Velocity

We consider now a phenomenon of great physical importance connected with dispersion. The derivative of the dispersion relation (13),

$$C(k) = \frac{dW}{dk} = W' \tag{15}$$

has the dimensions of velocity and is termed the *group velocity*. Consider a particular value of k and the corresponding value of $\omega = W(k)$ as reference values, and let the wavenumber be perturbed to $k + \Delta k$. The corresponding perturbed frequency may be approximated by the first two terms of a Taylor expansion as

$$\omega + \Delta\omega = W(k + \Delta k) \approx \omega + C\Delta k.$$

The phase θ_+ corresponding to this perturbed wavenumber is then (with α unchanged)

$$\theta_+ = (k + \Delta k)x - (\omega + \Delta\omega)t + \alpha = \theta + \Delta k(x - Ct),$$

where θ is the unperturbed phase of equation (11). If k is now perturbed so that it is decreased (instead of increased) by Δk to the value $k - \Delta k$, the corresponding phase (with α again unchanged) is

$$\theta_- = \theta - \Delta k(x - Ct).$$

The solution y_+ at wavenumber $k + \Delta k$ is

$$\begin{aligned}
y_+ &= y_0 \cos \theta_+ \\
&= y_0(\cos \theta \cos \Delta k(x - Ct) - \sin \theta \sin \Delta k(x - Ct)).
\end{aligned}$$

The solution y_- is defined as $y_0 \cos \theta_-$, with the same amplitude as y_+. The superposition of the two solutions gives

$$\begin{aligned}
y = y_+ + y_- &= y_0(\cos \theta_+ + \cos \theta_-) \\
&= 2y_0 \cos \theta \cos \Delta k(x - Ct).
\end{aligned}$$

This solution is sketched in Figure I.3 at $t = 0$.

This solution may be considered as one at the reference wavenumber and frequency with its amplitude modulated by the factor $\cos \Delta k(x - Ct)$. The solution has *beats*, corresponding to the changes in amplitude. The oscillation is bounded by the two curves

$$y = \pm 2y_0 \cos \Delta k(x - Ct), \tag{16}$$

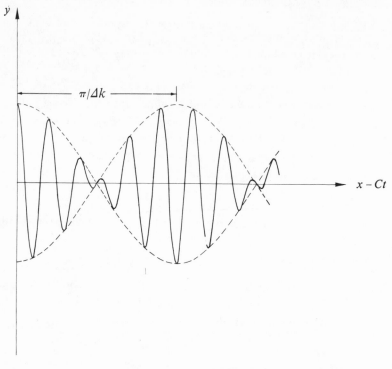

I.3. Solution with beats.

which appear as an envelope for the solution, or, more strictly for the family of solutions obtained by varying α.

The envelope or boundary curve (16) moves in space with the velocity C. Each lobe of the envelope, or boundary curve, may be interpreted as a group of waves, and the velocity C may be interpreted as the velocity of the group. In the nondispersive case we calculate from (14) and (15) that

$$C = c. \tag{17}$$

Thus the group velocity for nondispersive waves is the same as the phase velocity, a result consistent with our interpretation of nondispersive propagation as transmitting arbitrary signals unchanged. In the dispersive case the group velocity is different from the phase velocity and may be either greater than or less than (even of opposite sign to) the phase velocity.

Our results and interpretations of group velocity do not apply only to the particular envelope shape of (16) and Figure I.3. They apply also generally to wave packets with smooth envelope shapes subject to the condition that the extent of the packet (or lobe) is large compared with the wavelength $2\pi k^{-1}$ of the reference signal.

2d. Energy Conservation and Equipartition

At this stage we are working only with the assumed solution form (10) and dispersion relation (13), and we cannot define from them any energies. In most problems of physical wave propagation, energies are definable in terms of densities of energies per unit distance. Thus we usually have $E = T + V$, where E, T, and V are linear densities of total, kinetic, and potential energy, respectively.

Conservation of energy appears in the form of a relation

$$\frac{\partial E}{\partial t} + \frac{\partial S}{\partial x} = 0, \tag{18}$$

where S is an energy flux. This relation ensures that the total energy $\int E \, dx$ is constant if it is finite.

The average over a cycle used in our treatment of the oscillator is interpreted here as an average over the phase (over an interval of 2π). In most cases equipartition holds, and we have $\overline{T} = \overline{V}$. If the waves are nondispersive and are only of one family (no waves going in the opposite direction), the stronger local equipartition law $T = V$ may be shown to hold.

If the waves are only of one family, the energy flux S can usually be interpreted as EC. More generally, we can interpret the average $\overline{S}$ as $\overline{E}C$. It should be noted, however, that we may add an arbitrary constant to S or $\overline{S}$ without changing the relation (18). Although the identification of $\overline{S}$ as $\overline{E}C$ is a natural one, it does not determine S uniquely, and there are problems in which this identification is not direct. In most cases, however, we can visualize the energy density as moving with the group velocity. (See Chapter V.)

In some cases the undisturbed medium has infinite energy, and we cannot identify a conserved perturbation energy. Usually in these cases, the energy is replaced by a conserved *wave action* and the energy density by a wave action density. (See Chapters V and VI.)

In weakly dissipative systems, the amplitudes decrease in time with a factor $e^{-\beta t}$ and the energy densities with a factor $e^{-2\beta t}$. A group, or packet, of waves still moves with the group velocity and has an envelope shape that is a function of $x - Ct$. These damping factors are the factors as measured by an observer moving with the group velocity.

2e. Isotropy and Anisotropy

We digress at this point to describe how the concepts discussed above apply if the physical space is three-dimensional (or two-dimensional) instead of one-dimensional. Let x_1, x_2, x_3 be coordinates in a three-dimensional euclidean space, and let $\mathbf{x}$ be the coordinate vector. In place of (11), the phase is defined by

$$\theta = \mathbf{k} \cdot \mathbf{x} - \omega t + \alpha$$
$$= k_1 x_1 + k_2 x_2 + k_3 x_3 - \omega t + \alpha. \qquad (19)$$

The wavenumber is thus a vector quantity, normal to surfaces of constant phase θ. A velocity (in this case phase velocity) is not the most natural way to describe the motion of a surface through space, although it is frequently so used. In its place we use the *slowness vector* or *inverse phase velocity* given by

$$\mathbf{s} = \frac{\mathbf{k}}{\omega}$$

instead of (12). This has the property that $d\theta/dt$ is zero for an observer moving with a velocity $d\mathbf{x}/dt$ such that $\mathbf{s} \cdot d\mathbf{x}/dt = 1$; for an observer who stays on the same wavefront, $\theta = $ constant.

The dispersion relation (13) takes the form

$$\omega = W(\mathbf{k}).$$

A wave motion is termed *nondispersive* if the slowness vector $\mathbf{s} = \mathbf{k}/W(\mathbf{k})$ is independent of the magnitude of $\mathbf{k}$ and a function only of the direction of $\mathbf{k}$. If $\mathbf{k} = k\mathbf{n}$, where $\mathbf{n}$ is a unit vector in the direction of $\mathbf{k}$, the waves are nondispersive if $W = kc_n(\mathbf{n})$, where c_n is the inverse of the magnitude of $\mathbf{s}$.

Isotropy means "the same in all directions." A dispersion relation is *isotropic* if it is of the form

$$\omega = W(k) \tag{20}$$

independent of the direction **n** of **k**. Equation (20) appears to be the same as (13). There is a difference, however, and this will be taken up below.

The group velocity **C** is defined by

$$\mathbf{C}(\mathbf{k}) = W_{\mathbf{k}} = \nabla_{\mathbf{k}} W \tag{21}$$

in place of (15) and has components

$$\{\mathbf{C}\} = \left\{ \frac{\partial W}{\partial k_1}, \frac{\partial W}{\partial k_2}, \frac{\partial W}{\partial k_3} \right\}.$$

The trajectory of a point that moves with velocity

$$\frac{d\mathbf{x}_r}{dt} = \mathbf{C}$$

is termed a *ray*. The property (17) of nondispersive wave propagation becomes

$$\mathbf{C} \cdot \mathbf{s} = 1. \tag{22}$$

In this particular case the phase is constant on a ray.

The energy conservation law (18) generalizes to

$$\frac{\partial E}{\partial t} + \nabla \cdot \mathbf{S} = 0, \tag{23}$$

and the energy flux is a vector quantity. The arbitrary constant that may be added to S becomes an arbitrary solenoidal field that may be added to **S**.

2f. Standing Waves

We return to the one-dimensional case and ask what remains of the concept of anisotropy, or, in other words, what is the difference between the isotropic equation (20) and our one-dimensional relation (13). In (20) k is the magnitude of **k** and is always positive. In (13) k may have either sign. One-dimensional wave motion is isotropic if

$$W(-k) = W(k)$$

—that is, if W is an even function of k. This condition states, in essence, that waves moving to the left follow the same law as do waves moving to the right and that the problem in some sense is invariant with respect to a reflection of the x axis. We consider henceforth only cases in which this isotropy condition holds. We should note that a shift to a moving coordinate system (galilean transformation) destroys the isotropy property.

With this isotropy property, for each solution of the form

$$y = y_0 \cos (kx - \omega t + \alpha)$$

there is also a solution of the form

$$y = y_0 \cos (kx + \omega t + \alpha).$$

If these solutions are now superposed, the result is

$$y = 2y_0 \cos \omega t \cos (kx + \alpha). \tag{24}$$

With this solution there are fixed nodes at $kx + \alpha = (\pi/2) + \pi m$. The solution is purely oscillatory, with frequency ω. Such waves do not appear to be propagating and are termed *standing waves*.

When two waves of different families are superposed, as in the example that gives standing waves, the energy densities do not superpose. The reason is that the energy densities are quadratic in the amplitudes. The averaged energy densities do superpose, however, as the interaction terms have zero average. Some modification in the definition of an average may be needed.

2g. Nonpropagating and Convected Waves

The group velocity defined by (15) or (21) may be zero in certain special cases. In this case the waves are *nonpropagating*. There is no average energy flux $EC = S$, and a group does not move. This situation may always be attained in a trivial manner for a given mode by means of a shift to a coordinate system moving with the group velocity. In general, however, there is a natural coordinate system, usually with an isotropic dispersion relation, in which we can consider ourselves fixed.

A typical case that occurs in waveguide problems involves a dispersion relation of the form

$$\omega = W(k) = (\omega_c{}^2 + a^r k^2)^{1/2}, \tag{25}$$

where ω_c and a are constants. The group velocity in this case is

$$C = \frac{a^2 k}{(\omega_c{}^2 + a^2 k^2)^{1/2}}.$$

The group velocity approaches zero and the wave system becomes nonpropagating in the limit $k \to 0$. In this limit $\omega \to \omega_c$, which is termed the *cutoff frequency*. No frequencies less than the cutoff frequency can correspond to propagating waves. As the group velocity approaches zero, the phase velocity approaches infinity. They are related by $c \cdot C = a^2$. The wave system in this case becomes nondispersive in the limit $k \to \infty$, with $c = C = a$. (See Chapter II, Subsection 3d, for a discussion of the cutoff frequency.)

We next consider the case in which we can specify some independent variable in the form

$$y(x, t) = y_0 \cos (kx + \alpha), \tag{26}$$

independent of time and with any k. We now look at (26) as though it represented a wave. The dispersion relation is simply $\omega = W(k) \equiv 0$. The wave system has zero phase and group velocities, and is isotropic, nondispersive, and nonpropagating. As a wave system this is clearly trivial, and it is stretching the concept of a wave to apply the term "wave" to it.

However, if we shift to a coordinate system x' that is moving to the left with velocity v, the distribution (26) takes the form

$$y = y_0 \cos (kx' - kvt + \alpha).$$

This looks like the solution for a propagating wave with the nondispersive dispersion relation $W = kv$ and with $c = C = v$. The "propagation" is no longer isotropic, of course. Such a "wave" may be termed a *convected wave*. It should be considered to be convected rather than propagated, with the velocity v. Energy-conservation concepts do not generally apply to convected waves. The concept of propagation involves a motion of waves relative to a medium in some physical sense, and there is no such motion in this case. The expression (26) is not a solution of equations of motion.

Convected waves appear in a number of examples of physical interest, where some signal is convected by some stream. They are of particular interest usually when some nonlinear effect is important that modifies the otherwise quite trivial solution. An important physical example is to be found in the operation of klystron electron tubes.

2h. "Slip Frequency" and Dispersion

The foregoing discussion of dispersion was quite limited. The term nondispersive was used for a property of a dispersion relation valid over the entire range of wavenumbers. A dispersion relation may, in fact, possess a particular nondispersive property at one value of the wavenumber but not at others.

The frequency ω^T of a wavetrain with respect to a galilean coordinate system moving with the velocity $\mathbf{U}$ is

$$\omega^T = -\frac{d\theta}{dt} = -\frac{\partial\theta}{\partial t} - \mathbf{U} \cdot \nabla\theta$$

$$= \omega - \mathbf{k} \cdot \mathbf{U}.$$

This formula gives the magnitude of the Doppler frequency shift. For an observer moving with the group velocity $\mathbf{C}$, this frequency is termed the *slip frequency* and is

$$\omega^T(\mathbf{C}) = \omega - \mathbf{k} \cdot \mathbf{C} = \omega(1 - \mathbf{s} \cdot \mathbf{C}).$$

The property (22) of a dispersion relation nondispersive for all values of $\mathbf{k}$ is that the slip frequency is zero. In a general case, if property (22) held for a mode at a particular value of the wavenumber $\mathbf{k}$, these waves would be termed *nonslipping*. A wave rider constrained to a particular crest or wavefront could travel with the group velocity and thereby maintain his position with respect to the group only if this property held.

The term dispersion refers to the ability of a·wave packet or group to spread out or disperse and is a consequence of the nonconstancy of the group velocity $C(k)$ or $\mathbf{C}(\mathbf{k})$. This property, which permits changes in the overall shape of a wave packet, is measured by the derivative of the group velocity with respect to the wavenumber. For one-dimensional propagation the *dispersion* is defined

$$W''(k) = W_{kk} = \frac{dC}{dk}$$

in terms of the function (13). In the fully nondispersive case, for which (14) holds, this quantity is zero. If $W'' = 0$ for a particular value of k in the general case, waves at that value of k would be termed nondispersive.

In the three-dimensional case the derivative of $\mathbf{C}$ with respect to $\mathbf{k}$ is the symmetric *dispersion tensor*

$$W_{\mathbf{kk}} = \nabla_{\mathbf{k}}\nabla_{\mathbf{k}}W(\mathbf{k}) = \nabla_{\mathbf{k}}\mathbf{C}(\mathbf{k}).$$

For anisotropic wave propagation this tensor is not zero in general, even for waves of the type we have termed nondispersive (Alfvén waves are one exception). Since the tensor is symmetric, there exists a cartesian coordinate system with respect to which the tensor is diagonal, the diagonal elements being the three real eigenvalues of the tensor. The tensor may be singular, having one or two (but not three) of the eigenvalues zero, without the tensor being zero. In these cases the waves are partially nondispersive.

In the special case where

$$W(\mathbf{k}) = kf\left(\frac{\mathbf{k}}{k}\right)$$

the propagation is *nondispersive for all* $\mathbf{k}$ in the sense that the dispersion tensor is singular. In this case the nonslipping property (22) also holds. In general, the two properties (nondispersion and nonslipping) are not connected. In the isotropic case, with (20) holding, the dispersion tensor may be shown to have the components

$$(W_{\mathbf{kk}}) = \begin{pmatrix} W_{kk} & 0 & 0 \\ 0 & k^{-1}W_k & 0 \\ 0 & 0 & k^{-1}W_k \end{pmatrix}$$

in a coordinate system aligned with respect to $\mathbf{k}$. In the isotropic case (with nonpropagating waves excluded) the wave propagation is termed *nondispersive* at a given value of k if and only if $W_{kk} = 0$.

The two nonzero diagonal terms $k^{-1}W_k$ in the tensor above represent geometric dispersion, which is thus an inescapable effect in the isotropic case. The terms represent the change in the direction of the vector group velocity resulting from a change in the direction of **k**. The lateral divergence of rays from a source or virtual focus or the convergence of rays towards a focus corresponds to this effect.

2i. Kinematic Waves

A type of motion termed *kinematic wave motion* appears when an appropriate functional relation connects the density and flux of some physically conserved quantity (Lighthill and Whitham, 1955). Kinematic waves describe, approximately, flood surges in rivers and phenomena of traffic flow; the conserved quantities in these cases are quantity of water and number of automobiles. Kinematic waves are not at all waves in the sense generally used in this volume. Our interest lies in the analogy between kinematic waves and waves as treated elsewhere in this chapter.

With the motion of interest one-dimensional, with ρ the density of the conserved quantity, and with Q its flux, the conservation equation may be expressed

$$\frac{\partial \rho}{\partial t} + \frac{\partial Q}{\partial x} = 0.$$

The density and flux are assumed to be connected by a known functional relationship

$$Q = Q(\rho),$$

which is considered to be analogous to (13). The conservation equation may be restated as

$$\frac{\partial \rho}{\partial t} + Q' \frac{\partial \rho}{\partial x} = 0.$$

This has the solution that ρ is constant on trajectories (characteristics) of points moving with the "group velocity" $Q'(\rho)$. The analogy is completed by the observation that in ordinary wave propagation the "conservation equation" is

$$\frac{\partial k}{\partial t} + \frac{\partial \omega}{\partial x} = 0,$$

corresponding to phase or number of crests as the conserved quantity.

In the three-dimensional case the analogy fails. The flux $\mathbf{Q}$ is a vector, and the conservation equation is a scalar one:

$$\frac{\partial \rho}{\partial t} + \nabla \cdot \mathbf{Q} = 0.$$

The analogous equation for frequency and wavenumber is a vector one:

$$\frac{\partial \mathbf{k}}{\partial t} + \nabla \omega = 0.$$

The theory for kinematic waves can be extended if the functional relation is of the form

$$\mathbf{Q} = \mathbf{Q}(\rho)$$

[and cannot if $\rho = \rho(\mathbf{Q})$]. The "group velocity" is $\mathbf{Q}' = d\mathbf{Q}/d\rho$ in this case.

3. Simple One-dimensional Wave Equation

In this section we treat the simplest linear wave equation, with the time t and one space coordinate x as independent variables. Examples of some physical problems leading to this equation are discussed in Section 4.

The basic partial differential equation is

$$y_{tt} - a^2 y_{xx} = 0 \tag{27}$$

and is of the second order in both variables. The quantity a is a constant and has the dimensions of velocity. The equation is of the *hyperbolic* type, which means that it has associated with it particular lines in the (x, t) space termed characteristics. A more general discussion of many of the topics of this section will be found in Chapter III.

A generally fruitful approach to a new linear equation system that is suspected of having solutions of propagating-wave form is to assume this form and to seek the corresponding dispersion relation. To this end we substitute $y = y_0 e^{i\theta}$ into (27) with the operation of taking the real part in (10) understood and with the phase θ given by (11). The result is

$$y_0 e^{i\theta}(-\omega^2 + a^2 k^2) = 0.$$

The expression in the parentheses must be zero, and

$$\omega^2 = a^2 k^2.$$

This is the dispersion relation desired. In the form of (13) we can write this

$$W(k) = \pm ak.$$

The double sign $\pm$ indicates that there are two families of waves; note that the system is isotropic. It is also nondispersive, with the constant phase and group velocities given by

$$c = C = \pm a.$$

3a. General Solution

Since the wave system is nondispersive, the sinusoidal nature of the assumed solution is inessential. The general function $y = F(x - at)$ is a solution, as is also a general function of $x + at$. These may be superposed to give

$$y = F(x - at) + G(x + at). \tag{28}$$

The functions F and G must be twice differentiable but otherwise may be quite arbitrary.

The solution (28) is the general solution to (27). This statement means that any solution to (27) may be represented in the form (28). This representation is not completely unique, as we can add any constant to F and subtract it from G without change in the solution (28). To show that the solution (28) is a general one, we derive it in a direct manner. The new coordinates

$$\xi = x - at, \qquad \eta = x + at \tag{29}$$

are introduced in place of (x, t). Equation (27) becomes

$$-4a^2 y_{\xi\eta} = 0. \tag{30}$$

This equation is integrated with respect to η, with the result

$$y_\xi = F'(\xi). \tag{31}$$

The arbitrary "constant" in the integration with respect to η is really an arbitrary function of ξ, which we have labeled $F'(\xi)$. We next integrate (31) with respect to ξ and obtain

$$y = F(\xi) + G(\eta),$$

in a form identical with that of (28).

3b. Characteristics

For a partial differential equation in two independent variables, a characteristic is a path or line in the space of the variables along which the equation is equivalent to an ordinary differential equation in lower-order derivatives. Characteristics are important, for example, in the non-linear theory of gas dynamic motions. For (27) the characteristics are the two families of lines $\xi = const.$ and $\eta = const.$ The characteristics $\xi = const.$ are termed forward-running or right-running characteristics, as on them x increases with time. Correspondingly, the characteristics $\eta = const.$ are termed backward-running or left-running characteristics.

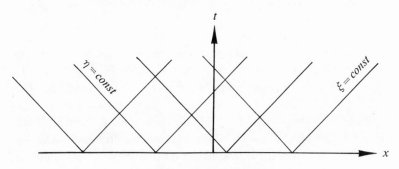

I.4. Characteristics of (27).

These are sketched in Figure I.4. Coordinates that are constant on characteristics are termed *characteristic coordinates*.

The appropriate analysis has essentially already been done in our derivation above of the general solution. We consider the quantities

$$2ay_\xi = ay_x - y_t, \qquad 2ay_\eta = ay_x + y_t \tag{32}$$

as distinct dependent variables. On right-running characteristics, from (29),

$$d\xi = dx - a\,dt = 0. \tag{33}$$

From (30) it follows that $2ay_\xi$ is independent of η, or that

$$d(2ay_\xi) = 0 \qquad \text{on } \xi = const. \tag{34}$$

Equation (34) is the ordinary differential equation equivalent to the partial differential equation on the characteristic defined by (33). In a similar manner, on left-running characteristics we have

$$d(2ay_\eta) = 0 \qquad \text{on } \eta = const.$$

These results appear to give us rather less information than that obtainable from the general solution, but the approach is important because it may be applied to a number of problems for which no general solutions are available. In such problems the analogs of (33) and (34) may be used as the basis of a numerical solution procedure. For example, the approach may be applied directly to the nonlinear equations (48) and (49) without further simplifying assumptions.

3c. Invariant Transformations

Equation (27), being a linear homogeneous equation, is invariant under multiplication of the dependent variable by a constant. The equation is also invariant under a number of transformations of the independent variables. The most general such transformation is that obtained by replacing ξ and η in (30) by monotonic differentiable functions of ξ and of η. These transformations include translations of the x scale, translations of the t scale, uniform expansions of both scales, and the Lorentz transformation.

The Lorentz transformation may be expressed in the form

$$\xi' = \left[\frac{1-M}{1+M}\right]^{1/2} \xi,$$

$$\eta' = \left[\frac{1+M}{1-M}\right]^{1/2} \eta,$$

or in terms of x and t in the form

$$x' = \frac{x + Mat}{(1 - M^2)^{1/2}},$$

$$at' = \frac{at + Mx}{(1 - M^2)^{1/2}},$$

(35)

where M is a dimensionless constant (a Mach number). The first of equations (35) shows the transformation is like a galilean transformation in being a shift to a moving coordinate system; the second shows the alteration of the time scale required to keep (27) invariant. The classical Lorentz transformation is of importance in special relativity and in electromagnetic theory, with a there being the speed of light. In the limit $M \to 0$, $a \to \infty$, with the coordinate system velocity Ma held fixed, the Lorentz transformation reduces to the galilean transformation.

The basic equation (27) is invariant under other operations. In particular, it is invariant under differentiation with respect to t or x. Thus, the quantities y_t and y_x satisfy (27) if y does, as higher derivatives do also. Any linear combination of y and its derivatives satisfies the equation. The particular linear combinations of y_t and y_x represented in (32) by y_ξ and y_η satisfy first-order partial differential equations as well, of course. With suitable conditions at infinity, the basic equation is invariant also with respect to integration along lines of fixed slope in the (x, t) space. Thus, for example, if $y(x, -\infty) = 0$, the quantity $\int_{-\infty}^{t} y(x, t^*) \, dt^*$ satisfies (27) if y does.

3d. Conservation Laws

A conservation law in a space of one (or more) dimensions plus time is a relation of the form (18) [or of (23)]. Equation (27) is itself in conservation form, as it may be written

$$\frac{\partial y_t}{\partial t} + \frac{\partial(-a^2 y_x)}{\partial x} = 0.$$

For *any* function $y(x, t)$ there are trivial conservation laws, of which the simplest is

$$\frac{\partial y_x}{\partial t} + \frac{\partial(-y_t)}{\partial x} = 0.$$

An important type of conservation law is one of the "energy" type, in terms of quantities identifiable as or analogous to physical energy density and flux. For (27) the simplest is

$$\frac{\partial}{\partial t}\left(\frac{1}{2}y_t^2 + \frac{1}{2}a^2 y_x^2\right) + \frac{\partial}{\partial x}(-a^2 y_t y_x) = 0. \tag{36}$$

Another is

$$\frac{\partial}{\partial t}(-y_t y_x) + \frac{\partial}{\partial x}\left(\frac{1}{2}y_t^2 + \frac{1}{2}a^2 y_x^2\right) = 0.$$

The freedom of generating new solutions by differentiation or integration must be kept in mind, as these generate new conservation laws.

With appropriate choices of the variable y, (36) is essentially the energy conservation equation (18) in a number of physical problems. To indicate the role played by (18) or by other conservation laws, we consider how the integral of E over x changes with time. This integral is expressed

$$\mathscr{E}(t) = \int_{x_1}^{x_2} E \, dx,$$

with the limits of integration assumed independent of t. The time derivative of $\mathscr{E}$ is

$$\frac{d\mathscr{E}}{dt} = \int_{x_1}^{x_2} \frac{\partial E}{\partial t} \, dx = -\int_{x_1}^{x_2} \frac{\partial S}{\partial x} \, dx = -S \Big|_{x_1}^{x_2}$$

$$= -S(x_2) + S(x_1).$$

In a number of problems either S vanishes at a finite boundary or approaches zero as the boundary approaches an infinite limit. In these problems $\mathscr{E}$ is constant. Conservation laws and conserved quantities are of use for a variety of reasons. Frequently the conservation laws are direct statements of the physical laws governing the problem. But they may also arise in the course of an approximation procedure, as in Chapters V, VI, and VIII.

3e. Initial and Boundary Conditions

In a typical initial-value problem, it is required to find the solution to the basic equation for $t > 0$, with $y(x, 0)$ and $y_t(x, 0)$ specified. Using the general solution (28), we can write

$$y(x, 0) = F(x) + G(x) \tag{37}$$

and

$$y_t(x, 0) = -a \, F'(x) + a \, G'(x). \tag{38}$$

In order to obtain the solution, we need to determine both functions F and G, and to determine them we need two initially specified distributions. We form an integral of (38),

$$-F(x) + G(x) = a^{-1} \int y_t(x, 0) \, dx. \tag{39}$$

The arbitrary constant in the integration has no effect on the solution. From (37) and (39) we may express the functions

$$F(x) = \tfrac{1}{2}y(x, 0) - \tfrac{1}{2}a^{-1} \int y_t(x, 0) \, dx$$

and

$$G(x) = \tfrac{1}{2}y(x, 0) + \tfrac{1}{2}a^{-1} \int y_t(x, 0) \, dx. \tag{40}$$

If the range of x is infinite, (40) gives the functions over the entire range of x and completes the solution. If the range of x is bounded, boundary conditions are needed. As an example we consider the range of x to be $x \geq 0$, with one boundary point at $x = 0$. In this case F and G are known from (40) only for positive values of their arguments. For $t > 0$, the argument of G is always positive and G is thus completely determined. On the other hand, the argument of F is negative over part of the range with $t > 0$, and is not determined there by (40). The missing information is obtained from a specification of y, y_t, or y_x as a function of time on the boundary $x = 0$. We assume that $y(0, t)$ is the quantity specified on the boundary and obtain

$$F(-at) = y(0, t) - G(at). \tag{41}$$

This relation gives us F for negative argument in terms of known functions and completes the solution. If the boundary condition is homogeneous in this case, with $y(0, t) \equiv 0$, the solution takes the form

$$y(x, t) = G(x + at) - G(at - x) \tag{42}$$

for $x < at$.

If the range of x is bounded both above and below, we require another boundary condition. If the range is $0 \leq x \leq l$, F and G are determined only for values of their arguments within this range. The boundary conditions at $x = 0$ provide information on F for negative argument through a relation like (41); if $y(l, t)$ is specified,

$$G(l + at) = y(l, t) - F(l - at) \tag{43}$$

provides information on G for values of its argument greater than l. If $y(t)$ is zero both at $x=0$ and $x=l$, we have from (41) and (43),

$$G(l+at) = -F(l-at) = G(-l+at).$$

The functions F and G are periodic of period $2l$ in this case, and the motion is periodic in time with period $2a^{-1}l$. The corresponding angular frequency ω_0 is

$$\omega_0 = \pi a l^{-1}. \tag{44}$$

3f. Normal Modes

We now examine further the solution obtained above with the homogeneous boundary conditions $y=0$ at $x=0$ and $x=l$. The function G is periodic of period $2l$, and may be expanded in a Fourier series:

$$G(x+at) = \frac{1}{2} \sum_{m=1}^{\infty} A_m \sin \frac{\pi m}{l} (x+at) - \frac{1}{4} B_0$$
$$- \frac{1}{2} \sum B_m \cos \frac{\pi m}{l} (x+at).$$

The solution (42) then takes the form

$$y(x, t) = \sum_{m=1}^{\infty} \sin \frac{\pi m x}{l} (A_m \cos m\, \omega_0 t + B_m \sin m\, \omega_0 t), \tag{45}$$

where ω_0 is given by (44).

In this form the solution may be interpreted in two ways. One, the solution is the same as a superposition of standing waves, each of the form of (24) but so arranged that the two boundaries are nodes. Two, the system is equivalent to an infinite number of simple linear oscillators, with frequencies $m\omega_0$, where $m = 1, 2, \ldots$, and with solutions of the form (3). In this interpretation the variable y for each value of m is the coefficient of $\sin(\pi m x l^{-1})$ in (45). These sine terms in x are termed *normal modes*. The frequency ω_0 is termed the *fundamental frequency* of the system and corresponds to waves of length $2l$. The length of the system is thus half the fundamental wavelength.

If the homogeneous boundary conditions are $y_x=0$ at $x=0$ and $x=l$, the results are essentially equivalent. The sines of $\pi m x l^{-1}$ are replaced by cosines. In fact, if y is given by (45), y_x is of this second type and satisfies $(y_x)_x = 0$ on the boundaries. The only difference is that an additional term constant in x, of the form $(A_0 + B_0 t)$, is permitted, one of zero frequency. This term is eliminated if the condition $\int_0^l y\,dx = 0$ is imposed.

Another possibility is that with the homogeneous boundary conditions $y = 0$ at $x = 0$ and $y_x = 0$ at $x = l$. In this case the solution is the same as that of (44) and (45), with l replaced by $2l$ but with only odd values of m appearing. The fundamental frequency ω_0 corresponds to waves of length $4l$, so l is one-quarter of the fundamental wavelength.

4. Physical Problems

4a. The Stretched String

With simplifying assumptions to linearize the problem, the theory of the stretched string leads to the basic equation (27). The string is assumed to be under constant tension F, with the lateral displacement of the string denoted by y and distance along the string by x. The inclination angle of the string is assumed small enough that its cosine may be approximated by one and its sine by its tangent y_x. The string is always in equilibrium in the x direction, and only the dynamics in the y direction are considered. The string is characterized by a uniform linear mass per unit length ρ. The lateral component of the tensile force is Fy_x. This provides a net lateral force on a segment of the string. This

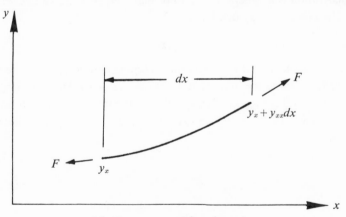

I.5. Forces on a string element.

lateral force (see Figure I.5) acting on a differential segment of the string of length dx is approximately

$$F(y_x + y_{xx}dx) - Fy_x = Fy_{xx}dx$$

and is proportional to the curvature of the string. Equating this force to the mass of the string segment $\rho\, dx$ times its lateral acceleration y_{tt} leads to (27)

$$y_{tt} - a^2 y_{xx} = 0,$$

with the wave speed a given by

$$a = \left(\frac{F}{\rho}\right)^{1/2}. \tag{46}$$

Waves in a stretched string are *transverse*, with the displacements at right angles to the x axis. Motion is also possible in the z direction, and displacements in the z direction obey the same law with the same wave speed. Consequently the system is degenerate, meaning that there are two distinct modes with the same dispersion relation. As with the linear spherical pendulum, the wave motion may be polarized; the motion is polarized in the y direction if $z \equiv 0$, or at $45°$ if $y \equiv z$. Circular polarization for a sinusoidal wave in one direction appears if

$$y + iz = y_0 e^{i\theta}.$$

Polarization is a general property of many types of transverse waves.

The kinetic energy density is

$$T = \tfrac{1}{2}\rho y_t^2$$

for motion only in the y direction. The potential energy may be interpreted as F times the increase in length of the string. Since the length of the string is the integral of $(1 + y_x^2)^{1/2}\, dx$, the increase in length per unit distance is $(1 + y_x^2)^{1/2} - 1 \approx \tfrac{1}{2}y_x^2$. Consequently, the potential energy density is

$$V = \tfrac{1}{2}Fy_x^2.$$

The energy flow across a point in the string is the lateral force $-Fy_x$ times the lateral velocity y_t, or

$$S = -Fy_t y_x.$$

Conservation of energy follows (36) with a factor ρ. The same results hold for motion in the z direction, with no interaction with the y motion.

The most familiar context in which vibrating stretched strings appear is in stringed musical instruments. In these cases the boundary conditions are $y = 0$ at $x = 0$, $x = l$. It is customary to consider the normal mode decomposition of the solution, with a mode at the fundamental frequency $\omega_0 = \pi a l^{-1}$ and with modes at the series of frequencies $2\omega_0$, $3\omega_0$, ... representing higher harmonics. Each mode may be considered as a separate linear oscillator with respect to initial conditions, polarization, energy, and so on.

In musical instruments, strings may be set into motion in three ways: by striking (e.g., piano), by plucking (guitar or harpsichord), or by bowing (violin). Bowing produces a continuing excitation, and our theory would have to be modified to describe the motion. Striking and plucking may be considered as establishing initial conditions. In all cases the motion of the string is imparted through one of the end supports to a sounding board, which then radiates sound waves. Since this process transfers energy from the vibrating string, the string acts as though it were damped. In general, the closer to one end the string is struck or plucked (or bowed), the more strongly the higher harmonics are excited relative to the fundamental. Higher harmonics are more strongly excited by striking than by plucking; this effect is counteracted (by design) in the piano by the soft felt hammers used, as these tend to damp the higher harmonics during the striking process.

A stretched membrane may show wave motion in two directions, say in the (x, z) space, with its displacement from the $y = 0$ plane described by a function $y(x, z, t)$. If we consider only motions that have no z variations (with $y_z = 0$) and let ρ be the density of the membrane per unit area and F its tensile stress or force in the x direction per unit depth, we obtain the same formulation as for the stretched string. The energy density is now per unit area; the motion is again transverse but is not degenerate and does not exhibit polarization.

4b. Acoustic Waves in Ducts

Consider the motion of a gas in a duct with cross-sectional area $A(x)$, under the approximation that the velocity u in the x direction and the thermodynamic functions of the gas are constant across the duct and are functions of x and t alone. Frictional effects are all neglected; other simplifying assumptions will be made as we proceed.

The time rate of change of the mass $\rho A\, dx$ in a segment of the duct of length dx is $\rho_t A\, dx$, where ρ is the gas density. This equals minus the net mass flow leaving the segment, which is $\{\rho u A + (\rho u A)_x\, dx\} - \rho u A$. This gives the continuity equation

$$\rho_t + u\rho_x + \rho(u_x + uA^{-1}A_x) = 0. \tag{47}$$

The first two terms in this expression are the time derivative of ρ as measured by an observer who moves with velocity u and thus with the fluid. For such an observer (with no dissipative effects), the pressure changes are related to the density changes by $dp = a^2\, d\rho$, where a^2 is the square of the velocity of sound in the gas. The continuity equation (47) may be rewritten

$$p_t + up_x + \rho a^2 u_x + \rho a^2 u(\log A)_x = 0. \tag{48}$$

The force on the parcel of gas in the duct segment of length dx is $-Ap_x\, dx$. The acceleration of the gas is the time derivative of the velocity as measured by the observer moving with the fluid, or $u_t + uu_x$. Equating the force to the mass $\rho A\, dx$ of the gas in the segment times this acceleration gives the momentum equation

$$\rho(u_t + uu_x) + p_x = 0. \tag{49}$$

We now assume that the motion is a small perturbation of a gas that is at rest under constant pressure. The velocity perturbation is u, and we denote the pressure perturbation Δp. The quantities ρ and a refer to the unperturbed distributions of density and sound speed. With higher-order terms dropped, the resulting linearized acoustic equations for the duct are

$$\Delta p_t + \rho a^2 u_x + \rho a^2(\log A)_x u = 0$$

and

$$\rho u_t + \Delta p_x = 0. \tag{50}$$

Equations (50) contain the two dependent variables u and Δp. Either variable may be eliminated to obtain a single equation for the other. If we eliminate Δp, the result is

$$u_{tt} - \rho^{-1}[\rho a^2 u_x + \rho a^2(\log A)_x u]_x = 0; \tag{51}$$

if we eliminate u, the result is

$$\Delta p_{tt} - a^2[\Delta p_{xx} + (\log A - \log \rho)_x \, \Delta p_x] = 0. \tag{52}$$

It is convenient in some problems to introduce a new dependent variable ξ defined by the relation

$$u = \xi_t$$

and interpreted as the displacement of a particle in the x direction. In terms of this variable, which also satisfies (51), the pressure perturbation is

$$\Delta p = -\rho a^2[\xi_x + (\log A)_x \xi].$$

Another new dependent variable that can be introduced is the mass-flow potential Ψ defined by

$$\Delta p = -\Psi_t,$$

in terms of which the velocity is

$$u = \rho^{-1}\Psi_x.$$

The variable Ψ obeys the same equation (52) as Δp.

The kinetic energy density per unit length is

$$T = \tfrac{1}{2}\rho A u^2. \tag{53}$$

The acoustic potential energy per unit length may be defined as

$$V = \frac{A \, \Delta p^2}{2\rho a^2}. \tag{54}$$

These densities satisfy the conservation law (18) in the form

$$\frac{\partial(T+V)}{\partial t} + \frac{\partial Au \, \Delta p}{\partial x} = 0; \tag{55}$$

this relation may be obtained directly from equations (50). The term $Au \, \Delta p$ is the rate at which work is done across a plane $x = const$.

In the simplest case, $\rho(x)$, $a(x)$, and $A(x)$ are all constants, the gas is uniform, and the duct area is constant. In this case (51) reduces to the basic equation (27)

$$u_{tt} - a^2 u_{xx} = 0,$$

with the wave speed the same as the sound speed. The variables Δp, ξ, and Ψ' satisfy the same equation. As a consequence of the restriction that $\rho(x)$ is constant, the quantity

$$\Phi = \rho^{-1}\Psi'$$

is a velocity potential, with $u = \Phi_x$. We shall consider henceforth only the case $\rho = const.$ and use Φ in place of Ψ'. The acoustic waves governed by (27) [or by (51) or (52)] are longitudinal, with the particle motion in the x direction. No degeneracy appears, nor does the phenomenon of polarization.

The energy equation (55) takes the form of the general conservation law (36) only if the variable used is ξ or Φ. A multiplicative factor is needed, of course, and is ρA if ξ is used or $\rho A a^{-2}$ if Φ is used.

One context in which these linearized waves in ducts appear is again that of musical instruments. The simplest example is that of an organ pipe, in which a column of air vibrates. The air column is excited at one end by a jet of air impinging on a sharp edge. At this end the boundary condition is essentially $\Delta p = 0$, $\Phi = 0$, $u_x = 0$, or $\xi_x = 0$. If the pipe is open (at the other end), the same boundary condition applies there, and the fundamental frequency is given by (44). If the pipe is closed (at the other end), the boundary condition there is $u = 0$, $\xi = 0$, $\Delta p_x = 0$, or $\Phi_x = 0$. In this case the fundamental is half that of equation (44), and only odd harmonics can be generated. A four-foot closed pipe and an eight-foot open pipe have the same fundamental frequency, and this frequency corresponds to a wavelength of sixteen feet.

Plane waves in a uniform gas in a three-dimensional space follow essentially the same analysis with A held constant and with x taken in the direction normal to the wavefronts. With the linearizing assumption, but without any restriction to plane waves, it may be shown that the governing equation is

$$\Phi_{tt} - a^2(\Phi_{xx} + \Phi_{yy} + \Phi_{zz}) = 0. \tag{56}$$

The appropriate energy densities are per unit volume and are given by (53) and (54) with u^2 replaced by $u^2 + v^2 + w^2$. The energy flux is Δp times the velocity vector.

4c. Elastic Waves

An isotropic elastic material is characterized by several elastic moduli. The two most familiar are the elastic modulus E for a rod and the shear elastic modulus G; they are related through the dimensionless Poisson ratio σ by

$$G = \frac{E}{2(1 + \sigma)}.$$

The rod modulus E is the ratio of stress to strain in a rod that is laterally unconstrained and with the lateral normal stresses zero. In a material constrained so that the lateral strains are zero, the corresponding modulus E' is different from (and larger than) E. Finally, there is an isotropic compression modulus K, the ratio between isotropic pressure and volume change. These moduli are related through

$$E' = K + \frac{4}{3}G = \frac{E(1 - \sigma)}{(1 + \sigma)(1 - 2\sigma)}, \qquad K = \frac{E}{3(1 - 2\sigma)}.$$

An isotropic elastic medium allows waves of various types. If the medium is unbounded, plane (local) longitudinal waves are possible and obey the basic equation (27) with

$$a^2 = c_l^2 = \frac{E'}{\rho}. \tag{57}$$

These waves are very much like acoustic waves and are characterized by changes in the density. There are also plane transverse waves obeying the law (27) with

$$a^2 = c_t^2 = \frac{G}{\rho}. \tag{58}$$

These shear waves produce no changes in the density, and they are transverse and exhibit the phenomenon of polarization. More general motions obeying (56) with either (57) or (58) are possible.

When the medium is not unbounded, the wave phenomena near the boundaries can be rather complicated. Longitudinal waves in a thin rod are, however, again simple. In a uniform thin rod, one-dimensional longitudinal waves exist that follow (27) with

$$a^2 = c_1^2 = \frac{E}{\rho}.$$

These waves have a lateral motion described by a modal structure and require the condition $kh \ll 1$, where h is the diameter of the rod. They are nondispersive only in the limit $kh \to 0$.

The elastic medium approaches a fluid in a limiting process in which $\sigma \to \frac{1}{2}$, $G \to 0$, $E \to 0$, $E' \to K$. In this limit, only the longitudinal waves with the speed given by (57) remain. This speed is the speed of sound in the fluid medium.

4d. Electromagnetic and Alfvén Waves

The presence of electric and magnetic fields greatly increases the variety of waves that can exist. Here we briefly look at two types. Maxwell's equations in a vacuum are

$$\nabla \times \mathbf{E} = -\frac{\partial \mathbf{B}}{\partial t}, \tag{59}$$

$$\nabla \times \mathbf{B} = \mu_0 \epsilon_0 \frac{\partial \mathbf{E}}{\partial t}, \tag{60}$$

$$\nabla \cdot \mathbf{E} = 0, \tag{61}$$

and

$$\nabla \cdot \mathbf{B} = 0, \tag{62}$$

where μ_0 and ϵ_0 are universal constants. For plane waves governed by these equations, consider the case in which $\mathbf{E}$ has only a z component and $\mathbf{B}$ has only a y component. With this restriction (61) and (62) are automatically satisfied if these components are functions only of x and t. Equations (59) and (60) then become

$$\frac{\partial E_z}{\partial x} = \frac{\partial B_y}{\partial t}$$

and

$$\frac{\partial B_y}{\partial x} = \mu_0 \epsilon_0 \frac{\partial E_z}{\partial t},$$

where the subscripts mean components and not differentiation. Elimination of B_y leads again to (27)

$$\frac{\partial^2 E_z}{\partial t^2} - c^2 \frac{\partial^2 E_z}{\partial x^2} = 0, \tag{63}$$

where c is the speed of light, given by

$$c = (\mu_0 \epsilon_0)^{-1/2}$$

in terms of the constants μ_0 and ϵ_0. The quantity B_y also satisfies (63) in place of E_z. The energy per unit volume is half electric and half magnetic and is conserved. The components E_y and $-B_z$ satisfy the same relations as do E_z and B_y. Electromagnetic waves are transverse, degenerate, and exhibit the phenomenon of polarization.

The other type of wave we shall consider is the Alfvén wave, which appears in a fluid of infinite conductivity (zero resistivity). The undisturbed fluid is at rest in a uniform magnetic field $\mathbf{B} = B\mathbf{i}$ in the x direction. Electric fields are all zero in a coordinate system in which the fluid is not moving. In place of (60) we have

$$\nabla \times \mathbf{B} = \mu_0 \mathbf{J},$$

with the assumption that the term in $\partial \mathbf{E}/\partial t$ may be neglected, and with $\mathbf{J}$ representing the current density. The magnetic field exerts a force per unit volume on the medium carrying the current equal to $\mathbf{J} \times \mathbf{B}$.

A consequence of the assumption that the conductivity is infinite is that the lines of magnetic force are frozen in the fluid. Thus if the medium is displaced laterally, and this lateral displacement varies with x, the magnetic force line initially in the x direction is tilted, and this process gives rise to a lateral perturbation component of the magnetic field.

For plane waves we assume the medium is displaced by a small amount δ in the y direction, with δ a function only of x and t. The perturbation magnetic field has only a y component, and is

$$B_y = B \frac{\partial \delta}{\partial x}.$$

The equation of motion is then

$$\frac{\partial^2 \delta}{\partial t^2} = \frac{1}{\rho \mu_0} \frac{\partial B_y}{\partial x} B = \frac{B^2}{\rho \mu_0} \frac{\partial^2 \delta}{\partial x^2}.$$

This is again the basic equation (27), with

$$a^2 = \frac{B^2}{\rho \mu_0}.$$

The assumption that the term in $\partial \mathbf{E}/\partial t$ may be neglected in the equation for $\nabla \times \mathbf{B}$ rests on the restriction that $a^2 \ll c^2$, that the Alfvén speed is small compared with that of light.

These Alfvén waves are transverse and have the type of degeneracy that gives polarization. There is a conceptual picture of the Alfvén waves that brings us back to the example of the stretched string. We may look at the magnetic lines of force as a distribution of stretched strings over an area. The tension is $\mu_0^{-1}B^2$ per unit cross-sectional area. The mass density is ρ per unit cross-sectional area. The Alfvén wave-speed formula (above) is then analogous to the formula (46) for the stretched string.

The limitation of δ to a function of x and t alone is inessential. It may be an arbitrary function of $x \pm at$ and z, and if the sound speed in the medium is negligibly small (cold plasma) compared with the Alfvén speed, it may be an arbitrary function of $x \pm at$, y, and z. In any case the dispersion tensor W_{kk} of Subsection 2h above is at least doubly singular (of rank 1); if the sound speed is negligible, it is zero. This unusual property of not having any geometric dispersion is consistent with the concept of a distribution of independent stretched strings, with propagation only along the strings.

5. Other Examples of Waves in a Single Space Coordinate

Besides the wave motions governed by the simple one-dimensional wave equation (27), there are a number of other types of one-dimensional wave motion. Some of these are closely related and are governed by equations quite similar to (27). Some are strongly dispersive and governed by quite different equations.

5a. Damped One-dimensional Wave Equation

Wave motion may be damped but otherwise arise in much the same way as does a motion following (27). The damping, or dissipation, may arise in several ways. A simple example is closely analogous to the damped linear oscillator of (6), governed by the equation

$$y_{tt} + 2\beta y_t - a^2 y_{xx} = 0.$$

While this equation does not yield directly a real dispersion relation, the transformation

$$y = e^{-\beta t} Y(x,\, t)$$

puts it in the form

$$Y_{tt} - \beta^2 Y - a^2 Y_{xx} = 0; \tag{64}$$

this equation does have the real dispersion relation

$$\omega^2 = a^2 k^2 - \beta^2. \tag{65}$$

This relation has several apparent anomalies. One of these is a low wavenumber cutoff at $k = k_c = a^{-1}\beta$, below which the solution is not oscillatory in time and no waves propagate. Another is that the apparent group velocity is equal to $a^2 k/\omega$, a quantity larger than a. The principal conclusion that should be made is that, unless $\beta \ll ak$, the propagation is strongly dissipative and the utility of the concept of a group velocity is lost (see Chapter II, Section 5).

If $\beta \ll ak$, the dissipation is weak, and the terms β^2 in (65) and $-\beta^2 Y$ in (64) may be neglected. The properties are essentially those of the corresponding nondissipative system (27), except for an exponential decay. This decay is given by the factor $e^{-\beta t}$ for the dependent variable y and by the factor $e^{-2\beta t}$ for the energy.

5b. Acoustic Waves in Ducts of Variable Area

We now consider the linearized equations for waves in ducts, with $\rho(x)$ and $a(x)$ constant but with $A(x)$ variable. The basic equation we use is (52), but written for the velocity potential Φ,

$$\Phi_{xx} + (\log A)_x \Phi_x - a^{-2}\Phi_{tt} = 0. \tag{66}$$

The transformation

$$\Phi = A^{-1/2}\mathcal{F}(x,\, t)$$

leads to the equation

$$\mathcal{F}_{xx} - B\mathcal{F} - a^{-2}\mathcal{F}_{tt} = 0 \tag{67}$$

for $\mathcal{F}$, where

$$B(x) = \frac{A_{xx}}{2A} - \frac{A_x^2}{4A^2} = \frac{1}{2A^{1/2}}\left[\frac{A_x}{A^{1/2}}\right]_x.$$

Equation (67) has plane wave solutions and a corresponding dispersion relation only if $B(x)$ is constant, or can be considered to be approximately constant.

One special case of interest is that in which $B(x)$ is identically zero. The area A is then proportional to the square of a linear function of x; we choose this function to be simply x^2, and thus set

$$A = x^2.$$

This case corresponds to waves in three-dimensional space with spherical symmetry about a point. Such motion, because of the spherical symmetry, behaves locally as though it were constrained in a cone of arbitrary small solid angle with vertex at the central point. The cross-sectional area of this cone is the solid angle times r^2, where r is the radial coordinate in a system of spherical coordinates. Our coordinate x is this radial coordinate.

Since the equation (67) with $B = 0$ is the same as (27), the same general solution is available. The general solution for equation (66) may thus be written

$$\Phi = \frac{1}{x} \, [F(x - at) + G(x + at)].$$

Except in the special cases where B is either zero or constant, we can obtain a simple wave solution only if the term in B is negligibly small. We assume a progressive wave solution but set

$$k^{-2} \, | \, B \, | \ll 1. \tag{68}$$

Equation (67) then yields the nondispersive dispersion relation

$$\omega^2 = a^2 k^2.$$

The general solution (28) applies, but with the restriction that the wavenumbers in the Fourier spectra of the functions F and G satisfy the restriction (68) to a good approximation. With this assumption we have the approximate solution

$$\Phi = A^{-1/2} \, [F(x - at) + G(x + at)].$$

This approximation is typical of a family of approximations often used in wave propagation problems, generally based upon the assumption that the wavenumbers involved are sufficiently large, that the waves are of sufficiently small wavelength. The Liouville-Green approximation (see Erdélyi, 1956, Chapter IV, and Fröman and Fröman, 1965),

often erroneously referred to as the WKB approximation, is of this
family and leads in wave problems to the familiar approximate theory
of geometric optics and the related theory of geometric acoustics.

6. Strongly Dispersive Waves

The linearized waves we have been considering have all been governed
either by the simple wave equation (27) or by an equation differing from
it only by the presence of terms with lower-order derivatives. In all
cases there is a characteristic velocity a that is the actual propagation
velocity, at least in some limit. There are various types of waves that are
strongly dispersive and follow equations quite different from (27). Perhaps the case most extensively studied is that of deep-water waves.
They may not even be describable by partial differential equations in
the variables x and t. We turn now to consider some other examples of
such waves.

6a. Elastically Supported Beam

Consider an infinite beam of mass density ρ per unit length, of bend-
ing stiffness EI, laterally elastically supported by a distributed spring of
spring constant K per unit length. We assume that the beam is under a
compressive axial load P. This axial load tends to buckle the beam (see
Figure I.6).

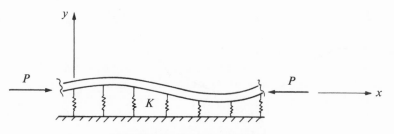

I.6. Elastically supported beam.

The lateral deflection of the beam is labeled $y(x, t)$. The lateral
acceleration y_{tt} times the mass $\rho \, dx$ of a section of the beam balances
the lateral force exerted on the section. This lateral force per unit
length is made up of three parts. One is the lateral force $- Ky$ exerted by

the elastic support. One, similar to that which appeared in the theory for the stretched string (with the sign changed, of course), is $-Py_{xx}$. The third may be obtained from the standard theory for beams and is $-EIy_{xxxx}$. Combining these, we obtain the differential equation

$$\rho y_{tt} + Ky + Py_{xx} + EIy_{xxxx} = 0$$

and the associated dispersion relation

$$\omega^2 = \rho^{-1}(K - Pk^2 + EIk^4).$$

We note first the special cases in which only a single restoring force exists. If $P = 0$ and $EI = 0$, the frequency is essentially the ω_0 of (1) for a linear oscillator and is constant; the waves are nonpropagating. If $K = 0$, $EI = 0$, and P is negative, we have (27) and (46) for a stretched string, with $F = -P$. If $K = 0$ and $P = 0$, we have waves in a simple unloaded beam. These are characterized by $C = 2c$.

If P is large enough, it is possible for ω^2 to be negative over a portion of the k scale. The condition that this not occur is

$$P \leq 2(KEI)^{1/2}. \tag{69}$$

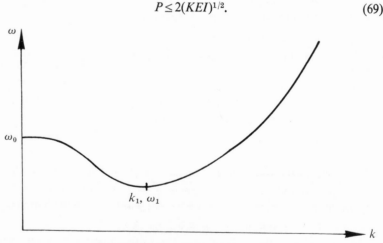

I.7. Dispersion relation for a beam.

With this condition met, the dispersion relation has the shape shown in Figure I.7. The waves are nonpropagating at $k = 0$, where $\omega = \omega_0 = (K/\rho)^{1/2}$. There is another nonpropagating point at

$$k^2 = k_1{}^2 = \frac{P}{2EI},$$

$$\omega^2 = \omega_1{}^2 = \frac{K}{\rho} - \frac{P^2}{4\rho EI} = \omega_0{}^2 - \frac{P^2}{4\rho EI}.$$

For $0 < k < k_1$ the group velocity is of opposite sign from the phase velocity. For $\omega_1 < \omega < \omega_0$ there are two wave modes with positive phase velocity, one with positive and one with negative group velocity. For a given positive group velocity that is sufficiently small there are three possible wave modes, one with positive and two with negative phase velocity.

If the inequality of (69) is not met, an infinitely long beam will buckle. If the beam, now more appropriately considered to be a column, is finite and satisfies boundary conditions $y = 0$ at $x = 0$, $x = l$, the buckling condition is that $\omega^2 < 0$ at $k = m\pi l^{-1}$ for some integer m. It is possible for the finite beam or column to be stable against buckling where a somewhat shorter column would buckle. Thus in this problem the dispersion relation contains not only information about wave propagation but also information about elastic buckling.

6b. Membrane Waveguide

In this example we consider a membrane. If the membrane has a uniform isotropic surface tension F and a mass density ρ per unit area, it has a characteristic speed a given by (46), $a^2 = F/\rho$. If $z(x, y, t)$ is the deflection of the membrane, the governing equation is

$$z_{xx} + z_{yy} - a^{-2}z_{tt} = 0.$$

The analogy of the membrane with the stretched string appears when we consider solutions with z independent of y. The physical problem we consider here is that of an infinite strip of width h, with the boundary conditions $z = 0$ on the lines $y = 0$ and $y = h$.

The method of separation of variables leads to solutions sinusoidal in y. The solution satisfying the boundary conditions is

$$z = e^{i\theta} \sin \frac{\pi m y}{h},$$

where m is a positive integer. These solutions in y are normal modes, with m the order of the mode. We define a *cutoff frequency* for the mth mode:

$$\omega_m = \frac{m\pi a}{h}.$$

When we substitute the assumed solution into the governing equation we obtain the mth dispersion relation:

$$\omega^2 = \omega_m{}^2 + a^2 k^2.$$

This dispersion relation has already been mentioned as (25), and we have discussed certain features associated with it. The group velocity c is always less than a and approaches a as $k \to \infty$. The waves are non-propagating if $\omega = \omega_m$. With a change of boundary conditions to $z_y = 0$, the same frequencies appear, plus a nondispersive mode corresponding to $m = 0$.

This type of behavior is characteristic of waveguide problems, including problems of electromagnetic waveguides. In the electromagnetic case the modal structure is more complicated, but each separate mode gives a dispersion relation of the form (25). The membrane waveguide gives the simplest model for studying propagation of this type. The only feature of more general waveguides that does not appear in this example is the possibility of degeneracy.

6c. De Broglie Waves

A general rule for dispersive waves is that the group velocity is physically of much greater importance than is the phase velocity (or slowness vector). One example that brings out this fact strongly is that of De Broglie waves, the solution of the Schrödinger equation for a free particle.

The time-dependent Schrödinger equation for a particle is

$$i\hbar \Psi_t = V(\mathbf{x})\Psi - \frac{\hbar^2}{2m} \nabla^2 \Psi,$$

where Ψ is the complex wave function and $2\pi\hbar$ is Planck's constant. With the potential energy V assumed constant, a plane wave solution of the form $\Psi = \Psi_0 e^{i\theta}$ is available, with the phase θ given by (19). The resulting dispersion relation is

$$\hbar\omega = \frac{\hbar^2 k^2}{2m} + V,$$

and the corresponding group velocity is

$$\mathbf{C} = \frac{\hbar\mathbf{k}}{m}.$$

Through the correspondence principle, $\hbar\omega$ is to be interpreted as the total energy, $\hbar^2 k^2/2m$ as the kinetic energy, $\hbar\mathbf{k}$ as the particle momentum, and hence the group velocity as the classical particle velocity. The group velocity thus has a specific physical interpretation.

The energy is arbitrary in the sense that an arbitrary constant may be added to all potential and total energies without changing the physical state. Thus, the frequency in the dispersion relation is arbitrary in the same sense, and this arbitrariness extends to the slowness vector $\mathbf{s} = \mathbf{k}/\omega$ or to any measure of the phase velocity. There is, however, no arbitrariness in the group velocity. The arbitrariness in the phase velocity indicates that it does not have any valid physical interpretation.

CHAPTER II

Linear Dispersive Waves

Stephen A. Thau

The theory of wave propagation is often introduced with the classical wave equation,

$$\phi_{tt} - c_0^2 \phi_{xx} = 0, \tag{1}$$

in which the constant c_0 has units of velocity and ϕ is a physical variable transmitted through the given medium in the form of a wave. Some familiar examples governed by (1) include transverse waves on a taut string, longitudinal waves in a thin elastic rod, plane sound waves in an unbounded compressible fluid, plane tidal waves on shallow water, and electrical waves in a transmission line with negligible resistance.

The general solution of (1) (see Chapter I, Section 3a),

$$\phi = f(x - c_0 t) + g(x + c_0 t), \tag{2}$$

shows that c_0 is the velocity with which the profiles $f(x)$ and $g(x)$ propagate in the positive and negative x directions, respectively, and that these profiles travel without change of shape. Thus, c_0 corresponds physically to the notion of a wave velocity.

On the other hand, the wave equation governs a relatively small number of examples and in several cases it provides only a first-order approximation that often must be refined. This is true in the afore-mentioned cases of waves in rods, water waves, and electrical waves. As shown in Chapter I, for a large group of wave motions, which are known as *dispersive*, more complicated equations than (1) describe the propagation, and solutions in the form of (2) do not occur. Even within the framework of linear theories of unidirectional wave propagation in homogeneous media, of which (1) is an example, it is found that wave profiles generally undergo marked changes in shape as they propagate.

Usually the initial waveform spreads, with an accompanying reduction of the wave elevation, and eventually the profile breaks down into an oscillatory pattern of crests and troughs, not unlike a sinusoidal wavetrain. Further, the concept of a wave velocity becomes more complicated and requires additional definitions for dispersive waves. The terms phase velocity, group velocity, signal velocity, energy propagation velocity, and front velocity are all encountered in the study of dispersive waves and they will all be discussed in this chapter.

1. Examples of Dispersive Waves. Phase and Group Velocities

The long list of dispersive wave motions embodies many well-known and important physical examples. Light waves in refractive media, seismic waves in the earth's crust, and underwater sound waves are familiar, although analytically complicated examples. Two others, for which the governing equations are simpler, are listed below in dimensionless form. These will be used later in this chapter to illustrate the theory of dispersive waves.

(i) Transverse waves along an elastically supported taut string are described by

$$\phi_{tt} - \phi_{xx} + \phi = 0. \tag{3}$$

This equation, known as the Klein-Gordon equation, also occurs in quantum mechanics.

(ii) Longitudinal waves in a thin elastic rod, including effects of lateral inertia, are described by

$$\phi_{tt} - \phi_{xx} - \phi_{xxtt} = 0. \tag{4}$$

Equation (4) is also known as the linear Boussinesq equation for shallow-water waves, after the man who first derived it. The same equation was derived by Love in his *Treatise on the Mathematical Theory of Elasticity* (1927), and so it is also known as Love's equation for waves in rods.

The theory of dispersive wave propagation is now introduced with a particular wave solution, namely, the simple harmonic wavetrains

$$\phi = A \exp\left[i(kx - \omega t)\right] + B \exp\left[i(kx + \omega t)\right], \tag{5}$$

which satisfy all linear equations of motion with constant coefficients in (x, t) space for suitable values of ω and k. However, what distinguishes dispersive waves is that the frequency ω becomes a *real, nonlinear* function W of the wavenumber k, for real k. These properties define linear dispersive wave equations in one space dimension. A more general discussion of dispersive waves is given in Chapter I, Section 2. Here we recapitulate some of the considerations of Chapter I for the special case of propagation in one dimension.

In any given problem the relation between W and k, which is called the dispersion relation for that problem, is found by direct substitution of one of the wavetrains in (5) into the governing equation. Thus, in the examples above the dispersion relations are obtained as

$$W = (k^2 + 1)^{1/2} \quad \text{and} \quad W = \frac{k}{(k^2 + 1)^{1/2}}.$$

Note that (5) satisfies the wave equation (1) but with the linear relation $W = kc_0$. Hence it is not a dispersive wave equation according to the definition of Chapter I. Also, the heat equation

$$\phi_{xx} - \phi_t = 0$$

affords an example where W would become a purely imaginary function of k, and for that reason it does not govern dispersive wave motion.

In the harmonic wavetrain in (5) with amplitude A, a point of fixed argument of the exponential, or fixed *phase*, is observed to travel in the positive x direction with the *phase velocity* (see Chapter I, Section 2)

$$c = c(k) = \frac{W(k)}{k}. \tag{6}$$

Similarly, the wavetrain with amplitude B moves in the opposite direction with the same phase velocity. For the wave equation, $c = c_0$ is independent of k. For dispersive waves, however, the relationship (6) indicates that *harmonic wavetrains with different frequencies propagate with different velocities*. Since an arbitrary waveform can be expressed as a superposition of harmonic wavetrains, this phenomenon explains why dispersive waves change in shape as they advance; the harmonic wavetrain components of a waveform at a given time, traveling with different speeds, become separated and at any later time will combine to form a different wave profile. In point of fact the dispersion of a waveform is caused by certain physical and/or geometrical characteristics of the medium in which the wave is generated. Consequently, instead of dispersive waves, it is perhaps more precise to speak of a *dispersive medium* or, where geometrical features alone cause the dispersion, a *dispersive geometry*. (See Chapter I, Subsection 2h.) Whatever the terminology, however, the meaning is given by (6).

Another velocity associated with the harmonic wavetrains (5) in dispersive media is the group velocity C defined in Chapter I as

$$C(k) = \frac{dW}{dk}, \tag{7}$$

which also depends on the wavenumber k. The group velocity is actually the most important velocity associated with dispersive waves, as it not only is the velocity of a given group of oscillations or "wavelets" in a wavetrain but also coincides with the velocity with which the energy in that group propagates. Moreover, as will be shown later, in a dispersive medium any initial disturbance is eventually broken up into a system of such groups.

In the previous examples, application of formulas (6) and (7) to the dispersion relations yields the phase and group velocities:

$$\text{(i)} \quad c = \left(1 + \frac{1}{k^2}\right)^{1/2}, \; C = \frac{k}{(k^2 + 1)^{1/2}};$$

$$\text{(ii)} \quad c = \frac{1}{(k^2 + 1)^{1/2}}, \; C = \frac{1}{(k^2 + 1)^{3/2}}.$$

It was indicated earlier that the frequency of a simple harmonic wavetrain in a dispersive medium should be real, which implies that the energy of the wave is conserved. However, dissipative mechanisms are always present to some degree in every real system. Mathematically, losses due to dissipation are manifested by the dispersion relation yielding complex or pure imaginary values of W corresponding to real values of k. The amplitude of a harmonic wavetrain then decays exponentially with time. In the case of a complex frequency with non-zero real part, the phase and group velocities can still be defined in terms of real quantities, but the group velocity in particular can lose its physical significance if the dissipation (i.e., the imaginary part of W) becomes appreciable. The general theory of waves in dispersive media would then require modifications. This subject will be discussed in Section 4.

On the other hand, complex values of k corresponding to real frequencies can occur in conservative systems with dispersion. One class of such problems involves the propagation of a number of modes that can exchange their energies. A mode with a complex wavenumber would thus lose its energy to (or absorb it from) other modes at the appropriate rate to conserve the total energy. This situation occurs for certain modes of stress wave propagation in elastic plates.

In addition, purely imaginary wavenumbers will occur at real frequencies in a conservative system that has a *cutoff frequency*. In example (i) above, k becomes imaginary for frequencies below $\omega = 1$, while in example (ii) k is imaginary for frequencies above $\omega = 1$. Such frequencies are called cutoff frequencies because the harmonic wavetrains (5) evanesce when k becomes imaginary; instead, they become standing waves whose vibration amplitude decays exponentially with distance. Physically, an initial harmonic disturbance at a frequency that yields an imaginary value of k will be dispersed into a system of wave groups having frequencies in the range where propagation is possible. An initial transient disturbance would be dispersed in a similar manner. However, a harmonic disturbance with a frequency in the range where k is real is transmitted directly through the medium with its group velocity. These results are illustrated in Section 3. In all conservative cases, the individual ripples that comprise a particular wave group advance within the group at the phase velocity, while the energy carried by the group of ripples propagates with the group velocity.

2. Harmonic Waves in Dispersive Media

The simplest illustration of a group, or packet, of oscillations that moves with a velocity different from that of the individual wavelets that comprise the group is the example presented by Stokes in 1876. It consists of the superposition of two harmonic wavetrains traveling in the same direction with slightly different wavenumbers and frequencies.

Letting $2\Delta k$ and $2\Delta\omega$ indicate these differences, respectively, we may express the two wavetrains (see Chapter I, Subsection 2c) as

$$\phi = A \exp \{i[(k + \Delta k)x - (\omega + \Delta\omega)t]\} + A \exp \{i[(k - \Delta k)x - (\omega - \Delta\omega)t]\}$$

which combine to form

$$\phi = 2A \cos (\Delta kx - \Delta\omega t) \exp [i(kx - \omega t)]. \tag{8}$$

This represents a harmonic wavetrain having wavenumber k and frequency ω but with its amplitude modulated by a cosine wave. Since $\Delta k/k$ and $\Delta\omega/\omega$ are assumed to be small quantities, the modulation is seen to be a slowly undulating waveform on both the x and t scales that envelops the more rapidly oscillating ripples. From the graph of the real part of (8) in Figure I.3, a wave packet can obviously be defined as the group of ripples contained between two successive nodes of the envelope. These ripples propagate with the phase velocity ω/k, whereas the velocity of the envelope is the group velocity, $C = \Delta\omega/\Delta k$. As $\Delta k \to 0$, C approaches the mathematical group velocity (7). In this same limit, however, the distance between successive nodes of the envelope becomes infinite, or, in other words, the wave groups disappear and a simple harmonic wavetrain is obtained. Nevertheless, this wavetrain can be regarded as an infinite group of ripples having the group velocity dW/dk.

The superposition of more than two harmonic wavetrains with nearly equal wavenumbers and frequencies is mathematically more complicated to analyze, but the group phenomenon can still be shown to occur (Havelock, 1914). Consider, for example, an infinite number of such wavetrains propagating in the positive x direction with average wavenumber k and average frequency $\overline{W} \approx W(\overline{k})$. Then ϕ can be expressed as

$$\phi = \frac{1}{\Delta k} \int_{\overline{k} - \Delta k}^{\overline{k} + \Delta k} A(k) \exp [i(kx - W(k)t)] \, dk, \tag{9}$$

where the amplitude A is taken to be a slowly varying function of k, and the factor $(\Delta k)^{-1}$ is introduced so that ϕ will remain nonzero as $\Delta k \to 0$.

The above integral can be evaluated approximately by expanding

W in a power series about k and retaining only the first two terms—in other words,

$$W \approx \overline{W} + (k - \overline{k})\overline{C},$$

where $\overline{C} = C(\overline{k})$ is the group velocity corresponding to the mean wavenumber. Then, with $k - \overline{k}$ replaced by κ and $A(k)$ approximated by $A(\overline{k}) = \overline{A}$, we obtain

$$\phi \approx \frac{\overline{A}}{\Delta k} \exp\left[i(\overline{k}x - \overline{W}t)\right] \int_{-\Delta k}^{\Delta k} \exp\left[i\kappa(x - \overline{C}t)\right] d\kappa$$

$$= 2\overline{A} \frac{\sin \Delta k(x - \overline{C}t)}{\Delta k(x - \overline{C}t)} \exp\left[i(\overline{k}x - \overline{W}t)\right]. \tag{10}$$

As in Stokes' example, the result is again found to be a wavetrain having the average wavenumber of the constituent simple harmonic wavetrains and having its amplitude modulated by a slowly varying waveform that envelops the ripples to form a system of wave packets. The packets travel with the group velocity $\overline{C}$, while the wavelets contained within propagate with the phase velocity $\overline{W}/\overline{k}$.

In the present example, the amplitude of the modulation is itself a varying function which, for fixed time, is seen to decrease in inverse proportion to x. Thus the maximum group is the one centered about the position $x = \overline{C}t$, where the amplitude is $2\overline{A}$. A plot of (10) is shown Figure II.1.

After a long time, the solution (10) will become invalid because the terms neglected in the above expansion of W, although being of order $(\Delta k)^2$, are not small when multiplied by t. In that event, the integral in (9) must be evaluated by a different method, known as the method of stationary phase, which is described in Section 3. This method shows that the series of wave packets in (10) will be dispersed into a more complicated system of wave groups whose amplitudes diminish at a rate proportional to t^{-1}—with one exception. As might be expected, the exceptional group still occurs at $x = \overline{C}t$, and its amplitude becomes proportional to $t^{-1/2}$.

3. Transient Waves in Dispersive Media

The propagation of harmonic wavetrains does not actually illustrate fully the phenomenon of wave dispersion. Both the simple harmonic

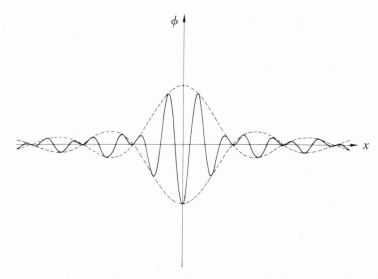

II.1. Wave packet (10) for moderate times.

waves in (5) and the harmonic wave packets formed by combining a finite number of different trains as in (8) propagate without change of shape throughout an extended medium. These are steady-state responses to harmonic sources that were "switched on" at $t = -\infty$, and the transient signal caused by the "switching on" has vanished by the time of our observation. On the other hand, the superposition of an infinite number of wavetrains as in (9) will be shown subsequently to be connected with the formation of an initial anharmonic disturbance at $t = 0$. At moderate times this initial disturbance is dispersed into the wave packets (10), while at later times [as explained following (10)] these packets will in turn be dispersed into more complicated groups of oscillations with continuously diminishing amplitude.

In this section we consider in full detail the dispersion of arbitrary initial waveforms. We also discuss the related initial-value problem of wave motion generated by a source function that is started at $t = 0$ in a previously undisturbed medium. The case of a time-harmonic source will be considered in order to show how a steady state wavetrain is actually developed behind a transient precursor in a dispersive medium. The procedures used here to analyze the transient dispersive wave problems include the Fourier transform method, the Laplace transform method, and a combined kinematic and energy approach developed by Whitham (Chapter V). The transform procedures lead to integral representations of the solution, which are then evaluated asymptotically by the stationary phase or steepest descent technique. The kinematic energy approach, on the other hand, yields essentially the same results directly from two simple conservation equations.

3a. Fourier Transform and the Method of Stationary Phase

Let us consider first the propagation of an initial disturbance in an extended dispersive medium occupying $-\infty < x < \infty$. A general solution for the wave motion can be expressed by a superposition of the simple harmonic wavetrains (5) over all possible wavenumbers—in other words,

$$\phi(x, t) = \int_{-\infty}^{\infty} \{A(k) \exp [i(kx - \omega t)] + B(k) \exp [i(kx + \omega t)]\} \, dk, \quad (11)$$

where the frequency ω is a known function $W(k)$ from the dispersion relation for the given medium and the amplitudes A and B are to be determined from the initial conditions.

$$\phi(x, 0) = f(x) \quad \text{and} \quad \phi_t(x, 0) = g(x).$$

We denote the Fourier transforms of the initial disturbances as $\bar{f}(k)$ and $\bar{g}(k)$, respectively, with

$$\bar{f}(k) = \int_{-\infty}^{\infty} f(x) \exp (-ikx) \, dx \quad (12)$$

and a similar equation to define $\bar{g}(k)$. The initial conditions, then, can be rewritten in terms of the inverse Fourier transforms

$$\phi(x, 0) = \frac{1}{2\pi} \int_{-\infty}^{\infty} \bar{f}(k) \exp (ikx) \, dk \quad (13a)$$

and

$$\phi_t(x, 0) = \frac{1}{2\pi} \int_{-\infty}^{\infty} \bar{g}(k) \exp (ikx) \, dk. \quad (13b)$$

By setting $t = 0$ in (11) and comparing the result to (13a), we obtain $A + B = \bar{f}/2\pi$. Similarly, after differentiating (11) with respect to time, taking $t = 0$, and using (13b), we find $B - A = \bar{g}/2\pi i\omega$. Hence

$$4\pi A = \bar{f} - \frac{\bar{g}}{i\omega} \quad \text{and} \quad 4\pi B = \bar{f} + \frac{\bar{g}}{i\omega},$$

which, when substituted into (11), completes the formal solution of the problem.

In the ensuing discussion of the solution, the initial value of ϕ_t is taken to be zero, so that $\bar{g} = 0$ and ϕ becomes

$$\phi(x, t) = \frac{1}{4\pi} \int_{-\infty}^{\infty} \bar{f}(k)\{\exp [i(kx - \omega t)] + \exp [i(kx + \omega t)]\} \, dk. \quad (14)$$

If ϕ satisfies the wave equation, then $\omega = kc$, and the integral in (14) is recognized as the sum of two inverse Fourier transforms of $\frac{1}{2}\bar{f}$ with arguments $x - ct$ and $x + ct$, respectively. Hence,

$$\phi(x, t) = \frac{1}{2}[f(x - ct) + f(x + ct)],$$

which is the well-known solution (2) of the wave equation satisfying the given initial conditions.

For a general dispersive medium, ω is a nonlinear function of k, and so the integral above usually cannot be evaluated exactly. For large times, however, we can obtain an asymptotic approximation for ϕ using the method of stationary phase. This is the method developed by Kelvin (1887) to treat just such an integral as in (14); it provides the detailed structure of the fully dispersed waveform.

The basic idea in Kelvin's approach is that the main contribution to the integral comes from small intervals centered about those values of k for which the phases of the exponentials are stationary. When t is large, the real and imaginary parts of the exponentials are rapidly oscillating functions along the k axis, except where the phases are constant or nearly constant as occurs in neighborhoods of stationary points. Thus the positive and negative contributions to the integral over most of the range of integration, where the crests and troughs are closely packed together, almost completely cancel one another. On the other hand, over regions where the phases are stationary the oscillations are much slower and the total effect is therefore much more substantial. From another point of view, the Fourier integral represents a super-position of an infinite number of harmonic wavetrains that initially produce the profile $f(x)$. After a short time these trains will have changed their positions with respect to one another, since each pro-pagates with its own distinct velocity. On physical grounds, however, it might be expected that enough wavelets are still in phase to produce a traveling waveform resembling $f(x)$; but after a long time the dis-tribution of phases at most stations x has become almost completely uniform. There are essentially an equal number of trains in and out of phase with each other so that their net effect is almost nil. Once again, though, there are exceptional positions where a preponderance of wavetrains have nearly equal phases and they combine to produce a relatively significant disturbance. This occurs where the phases of the exponentials are stationary.

The foregoing heuristic arguments can be justified mathematically, for it has been rigorously established that the method of stationary phase provides the leading term in the asymptotic expansion of Fourier integrals for large t (see, for example, Copson, 1965). In what follows, we present a brief review of the theory for obtaining the dominant term in the expansion of a generalized Fourier integral

$$I(t) = \int_{-a}^{b} V(k) \exp\left[it\Psi(k)\right] dk \tag{15}$$

for large positive values of t. The amplitude function V is assumed to be finite and differentiable over the range of integration.

First, consider the case where there are no points of stationary phase—that is, $d\Psi/dk \neq 0$ in $-a < k < b$. Equation (15) can be integrated by parts to yield

$$I(t) = \frac{i}{t}\left[-U(k)\exp\left(it\Psi\right)\Big|_{-a}^{b} + \int_{-a}^{b} U'(k)\exp\left(it\Psi\right) dk\right],$$

where primes indicate d/dk and $U = V/\Psi'$. The absolute value of the bracketed expression above is a bounded quantity; $|I(t)|$, therefore, is $O(1/t)$. This result is known as Riemann's lemma. It is still valid if a and b become infinite as in (14), provided that Ψ'' remains nonzero for all finite values of k and for $k \to \pm\infty$ as well. This would imply that Ψ' is a monotonically increasing or decreasing function that approaches $\pm\infty$ or $\mp\infty$ as $k \to \pm\infty$, respectively. However, in such a situation $|I|$ becomes much smaller than $1/t$, as can be shown by changing the integration variable in (15) to λ with $\lambda = \Psi'(k)$. Then we obtain

$$I(t) = \int_{-\infty}^{\infty} U(\lambda) \exp(it\lambda) \, d\lambda,$$

where $U(\lambda) = U(k(\lambda))$ is finite for real λ and $|U| \to 0$ as $|\lambda| \to \infty$ in order that the integral converge. By Jordan's lemma and Cauchy's formula, $I(t)$ can be evaluated as a series of residues at the poles of U in an upper complex λ plane, plus a sum of integrals along branch lines that extend from the branch points of U in the upper plane. Now, for $t \gg 1$, the dominant contribution comes from the singularity nearest the real axis of integration, $\lambda = \lambda_0 = \alpha + i\beta$, say with $\beta > 0$. If this singularity is a simple pole, then I is given by

$$I(t) \sim 2\pi i \exp[(i\alpha - \beta)t] \lim_{\lambda \to \lambda_0} [(\lambda - \lambda_0)U(\lambda)]. \qquad (16a)$$

For a branch point at λ_0, with $U(\lambda) = (\lambda - \lambda_0)^{-\delta} R(\lambda)$, where R is analytic at λ_0 and $0 < \delta < 1$, the branch line integral reduces to

$$I(t) \sim 2\pi R(\lambda_0) \frac{t^{\delta-1}}{\Gamma(\delta)} \exp\left[(i\alpha - \beta)t + \frac{i\pi\delta}{2}\right]. \qquad (16b)$$

Thus, in either case, $|I(t)|$ is $O(e^{-\beta t})$. Further details are discussed by Carrier, Krook, and Pearson (1966, pp. 255–257).

Let us now consider the integral (15) for the case in which $\Psi''(k) = 0$ at $k = k_0$, say, and $-a < k_0 < b$. We divide the range of integration into three segments as follows: from $-a$ to $k_0 - \epsilon$; from $k_0 - \epsilon$ to $k_0 + \epsilon$; and from $k_0 + \epsilon$ to b, where the constant 2ϵ is chosen to be small compared to $b + a$ but is large enough so that $\epsilon^2 t$ is still $O(t)$. From Riemann's lemma, then, the contributions from the first and third intervals are $O(1/t)$ and we may write

$$I(t) = J(t) + O\left(\frac{1}{t}\right) \qquad \text{with } J(t) = \int_{k_0-\epsilon}^{k_0+\epsilon} V \exp(it\Psi) \, dk \qquad (17)$$

The integral $J(t)$ can be evaluated for t large by expanding V and Ψ about k_0. This yields the asymptotic approximation

$$J(t) \sim V_0 e^{it\Psi_0} \int_{-\epsilon}^{\epsilon} \exp\left(i\frac{t}{2}\Psi_0''\kappa^2\right)d\kappa,$$

where κ has replaced $k - k_0$, $V_0 = V(k_0)$, $\Psi_0^{(n)} = \Psi^{(n)}(k_0)$, and we assume $\Psi_0'' \neq 0$. Recall that $\Psi_0' = 0$. Next, with $t|\Psi_0''|\kappa^2 = 2\rho$ and letting $\epsilon t^{1/2} \to \infty$ in the limits for ρ, we obtain

$$J(t) \sim \left(\frac{2\pi}{|\Psi_0'' t|}\right)^{1/2} V_0 \exp\left[i\left(\Psi_0 t \pm \frac{\pi}{4}\right)\right], \tag{18}$$

since

$$\int_{-\infty}^{\infty} \exp\left(\pm i\rho^2\right) d\rho = \pi^{1/2} \exp\left(\pm \frac{i\pi}{4}\right).$$

The $\pm$ signs correspond to $\Psi_0'' > 0$ and $\Psi_0'' < 0$, respectively.

Formula (18) is the stationary phase contribution. Note that it is $O(t^{-1/2})$ and hence is the dominant term in the asymptotic expansion of $I(t)$ for large t. The errors introduced in deriving (18) from (17), by truncating the expansion of V at V_0 and that of Ψ at Ψ_0'', and by letting the limits of integration for ρ become $\pm\infty$ instead of $\pm\epsilon t^{1/2}$, can be rigorously shown to be $O(1/t)$.

Based on the foregoing analysis, it follows that when $\Psi'' = 0$ at several points within the interval of integration, there are stationary phase contributions in the form of (18) from each one. Further, if a point of stationary phase coincides with a limit of integration, then the corresponding contribution is one-half that in (18). Finally, when Ψ''' is also zero at k_0, the analysis above becomes invalid and (18) cannot be used. The details for extracting the asymptotic behavior in this case will be presented shortly in connection with the dispersive wave solution (14).

Let us first apply the foregoing results to the example of Section 2, in which an infinite number of wavetrains with mean wavenumber $\bar{k}$ are superimposed (9). For small to moderate times, such that $(\Delta k)^2 t$ is small, the result is the system of wave packets (10). With t large, however, we must use the stationary phase method to evaluate ϕ. The stationary points occur for values of k satisfying

$$\Psi'' = \frac{x}{t} - W'(k) = 0,$$

or, equivalently, where

$$C(k) = x/t.$$

Now we have shown that the main wave packet is centered about the position $x = \bar{C}t$ with $\bar{C} = C(\bar{k})$ (Figure II.1). Hence, for $x/t = \bar{C}$, we have $k = \bar{k}$, which falls within the range of integration (9). The asymptotic value of the integral in (9) thus becomes

$$\phi \sim \frac{\bar{A}}{\Delta k} \left(\frac{2\pi}{t|C'|} \right)^{1/2} \exp\left[i(\bar{k}x - \bar{W}t \pm \pi/4)\right], \tag{19}$$

where we have used $\Psi'' = -W'' = C'$ so that the $\pm$ signs correspond to $W'' < 0$ and $W'' > 0$, repectively.

It is of interest to compare the long-time result above with the previous moderate-time expression (10). A wavetrain with wavenumber $\bar{k}$ and frequency $\bar{W}$ is still the dominant disturbance, but its amplitude is diminishing and is proportional to $t^{-1/2}$, and its phase has shifted by $\pm \pi/4$ radians. Furthermore, if we take $x = \bar{C}t \pm \Delta x$ instead of $x = \bar{C}t$, stationary phase points will still occur within the interval of integration, $\bar{k} - \Delta k$ to $\bar{k} + \Delta k$, provided that $\Delta x < \bar{C}' \Delta kt$, which becomes large as t increases. In other words, after sufficient time elapses, a point of stationary phase occurs for any *fixed value of* Δx, since $C = (x \pm \Delta x)/t$ will have a root, k, which approaches the point $\bar{k}$. Thus, eventually the wave packets (10) will completely disappear, and the wavetrain (19) will be observed in their place with its essentially constant wavenumber $\bar{k}$ and constant frequency $W(\bar{k})$. Nevertheless, for any *fixed* t, no matter how large, (19) still represents a finite group of oscillations that extends from $x = \bar{C}t - \frac{1}{2}\bar{C}' \Delta kt$ to $\bar{C}t + \frac{1}{2}\bar{C}' \Delta kt$. For values of x ahead of and behind this interval, no stationary phase points will occur in (9), and ϕ will therefore become $O(1/t)$ in these regions.

The general dispersive wave solution (14) for an arbitrary initial disturbance, $\phi = f(x)$, with Fourier transform $\bar{f}(k)$, can be similarly interpreted for large time. First we note that since $\phi(x, 0)$ is real, $\bar{f}(k) = \bar{f}^*(-k)$ from (13a). Furthermore, in the absence of dissipation ω is a real function of k, so that by noting that $\phi(x, t)$ is real and taking the complex conjugate of (14), we see that W must be either an even or an odd function of k. It then follows that (14) can be rewritten in the form

$$\phi = \frac{1}{2\pi} \text{Re} \int_0^\infty \bar{f}(k)\{\exp\ [i(kx - \omega t)] + \exp\ [i(kx + \omega t)]\}\ dk \qquad (20)$$

when $\bar{f}$ is even and "Re" means "real part of."

In (20), take $x > 0$, $m = x/t$ and assume that

$$W'(k) = C(k) = m$$

at $k = k_0(m) > 0$. Assume also that C is positive for positive k (negative group velocity is discussed later). Then the point of stationary phase for the first exponential is at k_0, while there is no point of stationary phase for the second exponential. Thus, in either case the stationary phase method yields the result

$$\phi(x,\ t) \sim \frac{\bar{f}(k_0)}{(2\pi t|C_0'|)^{1/2}}\ \cos\ (k_0 x - \omega_0 t \pm \pi/4), \qquad (21)$$

where, as before, the $\pm$ signs correspond to $C_0' \lessgtr 0$, respectively.

The expression (21) is a remarkable result. It says that any initial waveform is eventually dispersed into a harmonic wavetrain whose wavenumber k_0 and frequency $\omega_0 = W(k_0)$ depend on the ratio x/t. The amplitude of this wavetrain also varies as a function of x/t and attenuates as $t^{-1/2}$. However, within a neighborhood of a given station x and at a fixed time t, k_0, ω_0 and the amplitude factor $\bar{f}_0/|C_0'|^{1/2}$ appear as constants. We can check this by differentiating any of these functions with respect to x—for example:

$$\frac{\partial k_0}{\partial x} = \frac{1}{t}\frac{dk_0}{dm} = O\left(\frac{1}{t}\right). \qquad (22)$$

Hence, these quantities do not change appreciably over distances that are small compared to $C_0 t$.

In general the stationary phase solution (21) does not exhibit clearly delineated wave packets as discussed in Section 2. Nonetheless, for the purpose of interpreting the solution (21) we may arbitrarily define a wave group as the number of ripples occurring in an interval x to $x + \Delta x$. The interval length Δx should be large enough to include at least a few ripples, but it must be small compared to x. Then, based on

(22), such a group is comprised of essentially uniform harmonic waves. As before, the group is traveling with the group velocity C_0, and the ripples progress with speed ω_0/k_0. After a time Δt, the amplitude of the group has diminished by a factor $(1 + \Delta t/t)^{-1/2}$. If an observer attempted to follow a specific ripple at the phase velocity, he would soon leave the given group and find himself observing wavelets with a different wavenumber, frequency, and amplitude. On the other hand, if he traveled with velocity C_0, then he would continue to observe the wave-train (21); hence, C_0 is the group velocity with which fixed values of frequency and wavenumber will propagate. In Chapter V, Section 2, it is shown that the energy associated with the wavetrain (21) also propagates with the group velocity.

Initially, the waveform $f(x)$ is produced by a linear combination of harmonic wavetrains over the entire wavenumber spectrum. The wavetrain with wavenumber k_0, having amplitude $\bar{f}(k_0)/2\pi$, has no greater significance at $t=0$ than the other trains, particularly when F is a constant or nearly constant function. After a long time t, however, this very wavetrain appears by itself as the predominant disturbance at the station $x = C_0 t$. According to (21), its amplitude has changed and its phase has shifted. But the main point is that C_0 is precisely the group velocity of the wavetrain with wavenumber k_0. Hence this train, regarded as an infinitely long group of oscillations, has been traveling with its group velocity from time $t=0$ and will always be found at $x = C_0 t$. We also see, therefore, that no harmonic disturbance can travel faster than the maximum group velocity predicted by the given dispersion relation. The signal ahead of $x = C_{\max} t$ is either identically zero or is exponentially small.

A point of maximum group velocity (caustic), or for that matter of minimum group velocity, is of further interest because it brings up the situation in which $\Psi_0'' = - C_0' = 0$, so that the stationary phase solution (21) becomes incorrect as mentioned earlier. Let us suppose that C is maximum at $k = k_M$. Hence, at least in a neighborhood of k_M there are no points of stationary phase when $m\ (= x/t) > C_M$, while there are two such points for $m < C_M$. We again expand the phase function $\Psi = km - W(k)$ about k_M to evaluate the asymptotic behavior of the integral (20), only we now must include the cubic term. Also, to determine the nature of the solution for x near $C_M t$ as well as at $x = C_M t$, we take

$$m = C_M + \delta \quad \text{so that} \quad \delta = (x - C_M t)/t \qquad (23)$$

and $|\delta|$ is small. Thus we obtain

$$\Psi = m k_M - \omega_M + \delta(k - k_M) - \tfrac{1}{6} C_M''(k - k_M)^3 + O[(k - k_M)^4],$$

since $W_M' = C_M$ and $W_M'' = C_M' = 0$.

Now the analysis proceeds as before, with the result

$$\phi \sim \frac{\bar{f}(k_M)}{2\pi} \cos (k_M x - \omega_M t) \int_{-\epsilon}^{\epsilon} \cos [t(\delta \kappa - \tfrac{1}{6} C_M'' \kappa^3)] \, d\kappa,$$

where κ has replaced $k - k_M$. Next, we make the substitution

$$\kappa = \frac{\sigma p}{t^{1/3}}, \qquad \sigma = \left(\frac{2}{|C_M''|} \right)^{1/3},$$

and let $\epsilon t^{1/3} \to \infty$ in the limit for p. This yields

$$\phi \sim \frac{\sigma}{t^{1/3}} \bar{f}(k_M) Ai(\sigma \delta t^{2/3}) \cos (k_M x - \omega_M t), \qquad (24)$$

where the Airy function, $Ai(z)$, is defined by the integral representation

$$Ai(z) = \frac{1}{2\pi} \int_{-\infty}^{\infty} \cos (zp + \tfrac{1}{3} p^3) \, dp.$$

Recall that in our analysis $C_M'' < 0$. However, if a point where C is minimum had been treated, then $C_M'' > 0$ and the argument of the Airy function (24) becomes negative. Otherwise the result remains the same.

The Airy function is well known in applied mathematics (Abramowitz and Stegun, 1964, p. 446). For positive argument it behaves as an exponentially decaying function, whereas it oscillates sinusoidally for negative argument. These properties are exhibited by the asymptotic formulas,

$$Ai(z) \sim \frac{1}{2(\pi^2 z)^{1/4}} \exp (-\zeta) \qquad (25a)$$

and

$$Ai(-z) \sim \frac{1}{(\pi^2 z)^{1/4}} \cos \left(\zeta - \frac{\pi}{4} \right), \qquad (25b)$$

with $\zeta = \tfrac{2}{3} z^{3/2}$ and $z \to \infty$. Furthermore, at the origin ($\delta = 0$),

$$Ai(0) = \frac{3^{-2/3}}{\Gamma(\frac{2}{3})} \approx 0.355,$$

so that in a neighborhood of $x = C_M t$, the disturbances are $O(t^{-1/3})$, which are greater than the regular stationary phase contributions. These disturbances are sometimes called "Airy waves."

For values of $x > C_M t$ ($\delta > 0$), (25a) shows that the disturbances become exponentially small, which agrees with our previous result for the case where there are no points of stationary phase for the exponentials in (20). On the other hand, for $x < C_M t$, the Airy function in (24) becomes a cosine wave (25b) with velocity C_M and with wavenumber proportional to $t^{-1/3}$. This can be seen by substituting δ from (23) into (24). In other words, the Airy function acts as a slowly undulating amplitude modulation for the wavetrain in (24) that travels with the group velocity C_M. This is just as if two trains had been superimposed to form the Stokes wave packets in (8)—which is precisely what has happened, since there are two neighboring points of stationary phase about $k = k_M$ for values of m less than the maximum group velocity C_M and hence there are two wavetrains of the type (21) with wavenumbers differing slightly from k_M that must be superimposed.

In point of fact, the Airy waves can be rigorously matched with the asymptotic solution for the disturbances immediately ahead of and behind them. This matching in a specific example will be illustrated in Subsection 3c. First, however, we shall discuss the Laplace transform method for analyzing transient, dispersive wave propagation.

3b. Laplace Transform and the Method of Steepest Descent

The Laplace transform of a source function $P(t)$ at $x = 0$ in a dispersive medium is given by

$$\hat{P}(s) = \int_0^\infty P(t) \exp(-st)\, dt,$$

in which s is a complex number with positive real part and $\hat{P}(s)$ is to be interpreted as the analytic continuation of the integral in the entire s plane. It is assumed that $P(t) = 0$ for $t < 0$ and that there are no initial disturbances. The waves at any later time and position $x > 0$ can then be represented by the inverse Laplace transform integral,

$$\phi(x, t) = \frac{1}{2\pi i} \int_L \hat{P}(s) \exp [st - \gamma(s)x] \, ds \equiv \frac{1}{2\pi i} \int_L \hat{P}(s) \exp [xh(s)] \, ds, \qquad (26)$$

where the line L is a vertical line lying to the right of all singularities of $\hat{P}$ and γ, and the integration is from $-i\infty$ to $i\infty$. The function $\gamma(s)$ can be found from the dispersion relation for the given medium, $\omega = W(k)$. Setting $\omega = is$ and solving for $k = k(is)$, we define γ as

$$\gamma(s) = \pm ik(is),$$

where the choice of sign is dictated by the condition that the real part of γ should be positive when s is a real, positive number. The function $h(s)$ is defined by

$$h(s) = \frac{s}{m} - \gamma(s).$$

For example, in case (i) of Section 1

$$W = (k^2 + 1)^{1/2} \quad \text{so that} \quad \gamma(s) = (s^2 + 1)^{1/2},$$

while in case (ii)

$$W = \frac{k}{(k^2 + 1)^{1/2}} \quad \text{so that} \quad \gamma(s) = \frac{s}{(s^2 + 1)^{1/2}}.$$

These same results can be deduced by applying the Laplace transform directly to the governing equation of motion and solving the resultant ordinary differential equation for $\bar{\phi}$, subject to the boundary condition at $x = 0$, $\bar{\phi} = \hat{P}(s)$.

As in the case of the Fourier integral (14), the integration in (26) usually cannot be performed exactly; but for large values of x and t, an asymptotic approximation can be obtained. In the present case, however, the integration is along a curve in a complex plane, and the argument of the exponential is complex; therefore, the stationary phase method is not applicable. Instead, we employ a more general procedure known as the method of steepest descent, or the saddle point method.

In this method we are still concerned with the points where the argument of the exponential has a zero derivative, although the motivation is different. First, we shall seek a different path of integration L', with the following properties:

(1) L' passes through the point s_0 where the derivative of $h(s)$ vanishes—that is, where $\gamma'(s) = 1/m$;

(2) the imaginary part of $h(s)$ is everywhere constant on L', being equal to the imaginary part of $h(s_0) = h_0$,

(3) the real part of $h(s)$ attains an absolute maximum at s_0 along L'; and

(4) $\int_{L'}$ is equivalent to $\int_{L}$ or differs from it by a finite residue contribution.

In what follows we assume $h(s)$ is an analytic function of s given by $u + iv$. We know from the Cauchy-Riemann conditions that if u is to be maximum at s_0, then $h'(s_0)$ must be zero. However, there are actually two contours of constant v through s_0. These are orthogonal at s_0, and $u_0 [= u(s_0)]$ will be a maximum on one and a minimum on the other. Thus, s_0 is a saddle point or "col" for u. We shall prove these results by determining the equations of the tangent lines to the curves through s_0 on which $v = v_0$.

Take $s - s_0 = \rho \exp(i\theta)$; in a neighborhood of $\rho = 0$ we have

$$h(s) - h(s_0) \approx \tfrac{1}{2} h_0'' \rho^2 \exp(2i\theta), \tag{27}$$

so that by letting $h_0'' = -\gamma_0'' = |h_0''| e^{i\mu}$ and setting $v = v_0$, we obtain

$$\sin(2\theta + \mu) = 0,$$

whose distinct solutions in the s plane become

$$\theta = \theta_1 = -\frac{\mu}{2} \quad \text{and} \quad \theta = \theta_2 = -\frac{\mu}{2} + \frac{\pi}{2}.$$

These are the slope angles of the tangent lines. The real part of (27), evaluated along these lines, now becomes

$$u - u_0 = \pm \tfrac{1}{2} |h_0''| \rho^2,$$

corresponding to $\theta_{1,2}$, respectively. Along $\theta = \theta_2$, therefore, u_0 is a maximum; while along $\theta = \theta_1$, u_0 is a minimum.

Having found the curve L', we can then argue that the dominant contribution to the integral (26) along L' comes from a neighborhood of s_0, in other words, it comes from integrating along the tangent line to L' at s_0, from $s_0 - \epsilon$ to $s_0 + \epsilon$. This is because the real part of h is decreasing at an exponential rate along L' in either direction from s_0. Further, with the imaginary part of h constant, no cancellations occur from a rapidly oscillating integrand. Consequently, the contribution from the remainder of the integration path is of exponentially smaller order than that from the vicinity of s_0.

The rigorous mathematical justification for this method is given by Copson (1965), who uses the terms "saddle point method" when the tangent line is used at s_0 and "steepest descent method" when the entire curve L' is used. Here we shall proceed to evaluate the leading term in the asymptotic expansion of the waveform (26) for large x and t. The steps parallel those used in the stationary phase method. First $\hat{P}(s)$ and $h(s)$ are expanded about s_0, the former to order unity and the latter to order $(s - s_0)^2$. Thus,

$$\phi \sim \frac{\hat{P}(s_0)}{2\pi i} \exp(xh_0) \int_{s_0-\epsilon}^{s_0+\epsilon} \exp\left[\frac{x}{2} h_0''(s - s_0)^2\right] ds,$$

where, as stated above, the path of integration is the tangent line $\theta = \theta_2$ and so ϵ is a complex number, given by $|\epsilon| \exp[i(\pi - \mu)/2]$ with $|\epsilon|$ small. Next, the integration variable s is changed to a real variable λ by the substitution

$$\lambda = \left(\frac{x}{2}|h_0''|\right)^{1/2} (s - s_0) \exp[i(\pi - \mu)/2],$$

and the limits of integration for λ are taken to $\pm\infty$, since x is large. Then ϕ becomes

$$\phi(x, t) \sim \frac{\hat{P}(s_0)}{(2\pi x|h_0''|)^{1/2}} \exp\left[s_0 t - \gamma(s_0)x - \frac{i\mu}{2}\right], \qquad (28)$$

where we have used the formulas

$$\int_{-\infty}^{\infty} \exp(-\lambda^2)\, d\lambda = \pi^{1/2}$$

and $xh(s) = st - \gamma(s)x$.

This completes the derivation of the saddle point expansion. The entire asymptotic series of inverse powers of x can also be extracted by retaining all the terms in the expansion of $\hat{P}$ and h about s_0. Details are given by Copson (1965), pages 63–106, in several examples.

In the next two sections we shall illustrate the use of this method in analyzing specific problems of dispersive wave motion. When m is such that there are real points of stationary phase satisfying $m = C(k)$, then the solution (28) becomes equivalent to the stationary phase result (21). On the other hand—and this is a particular advantage of the saddle point method—when there are no real roots of $C(k) = m$, then (28) yields directly the exponentially decaying signal discussed previously. That expression is not easy to determine explicitly from the Fourier integral (20).

3c. Dispersion of a Delta Pulse in an Elastic Rod

The equation of motion governing longitudinal wave propagation in a thin elastic rod, which includes the effects of lateral inertia, was presented in example (ii) in Section 1. The dispersion relation and group velocity for this equation are given by

$$W = \frac{k}{(1 + k^2)^{1/2}}, \qquad C = \frac{1}{(1 + k^2)^{3/2}}. \tag{29}$$

It is observed that the maximum group velocity C_M is equal to unity and it occurs at $k = 0$. Therefore, for $m = x/t > 1$, we expect an exponentially small signal, and in the vicinity of $m = 1$ there should be Airy waves with amplitude of order $t^{-1/3}$. Trailing these will be harmonic waves of the form (21) with amplitude proportional to $t^{-1/2}$. The solution will be derived below for the case of an infinite rod $-\infty < x < \infty$ with an initial disturbance

$$\phi(x, 0) = \delta(x), \qquad \phi_t(x, 0) = 0,$$

where $\delta(x)$ is the Dirac delta function. The results are equally valid for gravity waves on the surface of shallow water according to the Boussinesq approximation.

By symmetry it suffices to treat the positive half of the rod, $x > 0$.

Setting $C = m$ to determine the wavenumbers for the predominant disturbance at station x, we obtain from (29)

$$k = k_0 = (m^{-2/3} - 1)^{1/2},$$

so that

$$\omega_0 = (1 - m^{2/3})^{1/2} \quad \text{and} \quad C_0' = -3m^{4/3}(1 - m^{2/3})^{1/2}.$$

As anticipated, k_0 is real only for $m < 1$. Thus, substitution of the above relationships into (21), with $\bar{f} = 1$, yields for $m < 1$

$$\phi \sim \frac{m^{-2/3}}{(6\pi t)^{1/2}(1 - m^{2/3})^{1/4}} \cos\left(k_0 x - \omega_0 t + \frac{\pi}{4}\right)$$

$$= \frac{m^{-2/3}}{(6\pi t)^{1/2}(1 - m^{2/3})^{1/4}} \cos\left[t(1 - m^{2/3})^{3/2} - \frac{\pi}{4}\right]. \qquad (30)$$

As $m \to C_M = 1$, the expression above becomes invalid and the Airy wave solution (24) should be used instead. However, since in this problem the Airy phase occurs at $k = 0$, which is at an endpoint of integration in (20), we have to take one-half of the result in (24). Thus, for $m = 1 + \delta$, we obtain

$$\phi \sim \frac{1}{2}\left(\frac{2}{3t}\right)^{1/3} Ai\left[\left(\frac{2t^2}{3}\right)^{1/3}\delta\right], \qquad (31)$$

since $C_0'' = -3$ and $k_0 = \omega_0 = 0$ at $m = 1$.

To complete the solution, we need the asymptotic behavior of the wave motion for $m > 1$. This is most conveniently derived by the Laplace transform and steepest descent approach.

If the Laplace transform is applied to (20), we obtain

$$\phi(x, s) = \frac{s}{\pi} \operatorname{Re} \int_0^\infty \bar{f}(k) \frac{\exp(ikx)}{s^2 + \omega^2} \, dk,$$

which, for the problem at hand, becomes

$$\phi(x, s) = \frac{s}{\pi} \operatorname{Re} \int_0^\infty \frac{k^2 + 1}{k^2(1 + s^2) + s^2} \exp(ikx) \, dk$$

$$= \frac{1}{2}(1 + s^2)^{-3/2} \exp\left[-\frac{sx}{(1 + s^2)^{1/2}}\right] \quad \text{for } x > 0.$$

Hence an equivalent representation of the solution is given by the inverse Laplace transform

$$\phi(x,\ t) = \frac{1}{4\pi i} \int_L (1 + s^2)^{-3/2} \exp\left[st - \frac{sx}{(1 + s^2)^{1/2}} \right] ds.$$

A branch cut is introduced in the s plane between the branch points at

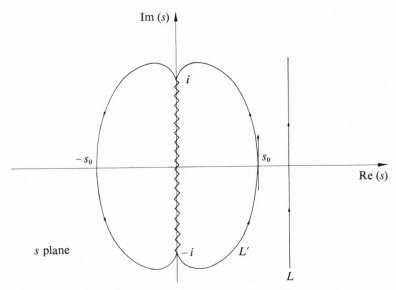

II.2. Path of integration for the Laplace transform representation of the dispersion of a pulse in an elastic rod.

$\pm i$, as shown in Figure II.2, to render the function $(1 + s^2)^{1/2}$ single-valued. Next, the cols are found at

$$s = \pm s_0 = \pm (m^{2/3} - 1)^{1/2}$$

which are real for $m > 1$. The path L' of steepest descent through s_0 is shown in Figure II.2. It has a vertical tangent at s_0 and is equivalent to the original path L. Thus we may apply the saddle point method to integrate on L' and obtain from (28)

$$\phi \sim \frac{m^{-1/6}}{2(6\pi x)^{1/2}(m^{2/3}-1)^{1/4}} \exp\left[s_0 t - \frac{s_0 x}{(1+s_0^2)^{1/2}}\right]$$

$$= \frac{m^{-2/3}}{2(6\pi t)^{1/2}(m^{2/3}-1)^{1/4}} \exp\left[-t(m^{2/3}-1)^{3/2}\right] \tag{32}$$

for $m > 1$.

The disturbances for $m < 1$ (30) and $m > 1$ (32) can be matched with the Airy waves near $m = 1$ (31). Setting $m = 1 \pm |\delta|$ in (32) and (30), respectively, and expanding the resultant expressions about $\delta = 0$ yields the leading terms

$$\phi \to 2^{-3/4}(3|\delta|)^{-1/4}(\pi t)^{-1/2} \cos\left[(\tfrac{2}{3}|\delta|)^{3/2}t - \frac{\pi}{4}\right] \qquad (\delta < 0) \tag{33a}$$

and

$$\phi \to 2^{-7/4}(3\delta)^{-1/4}(\pi t)^{-1/2} \exp\left[-(\tfrac{2}{3}\delta)^{3/2}t\right]. \tag{33b}$$

These agree exactly with the asymptotic expansions of (31) for $t^{2/3} \cdot \delta \to \pm\infty$, respectively. The latter can be deduced from (25).

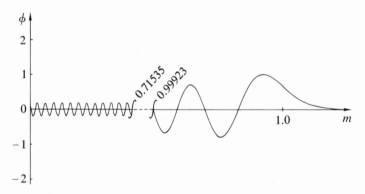

II.3. Dispersed waveform corresponding to (30), (31), and (32) for $t = 10^6$, starting from $m = 0.71475$.

A numerical plot of the completely dispersed waveform (30), (31) and (32) is presented in Figure II.3 for $t = 10^6$ and from $m = 0.71475$ to 1.00038. Bear in mind that the elementary theory for waves in a rod is the wave equation (1) that would predict a constant waveform $\phi = \frac{1}{2}\delta(x - t)$ advancing with unit velocity in the positive x direction.

Here, with dispersion taken into account, the result is much different. A finite harmonic wave occurs for all $x < t$, ever diminishing in amplitude as it spreads out over a larger distance. The predominate disturbance is found near $x = t$ and it propagates with unit velocity. However, it is preceded by an exponentially decaying precursor.

3d. Generation of Harmonic Waves on a Dispersive String

The second example of transient wave motion in a dispersive medium will employ the linear Klein-Gordon equation (3) of example (i) in Section 1, which governs the propagation of waves on an elastically supported string. We shall consider a semi-infinite string $x > 0$, initially at rest and subjected at $x = 0$ to the unceasing sinusoidal disturbance

$$\phi(0, t) = \exp(i\Omega t) \qquad \text{for } 0 \leq t.$$

We apply the Laplace transform to (3) and the above boundary condition and readily deduce the solution for ϕ in the form

$$\phi(x, t) = \frac{1}{2\pi i} \int_L \exp[st - (s^2 + 1)^{1/2}x] \frac{ds}{s - i\Omega}, \tag{34}$$

where the integrand is made single-valued by branch cuts in the left-hand s-plane from $-i$ to i. The pole at $i\Omega$ can be located either above (as shown in Figure II.4) or below the branch point $s = i$, but we shall take Ω to be positive.

Before evaluating the inverse Laplace transform (34), let us examine the dispersion relation and group velocity for the equation of motion (3). These were found to be

$$W = (1 + k^2)^{1/2} \quad \text{and} \quad C = \frac{k}{(1 + k^2)^{1/2}}, \tag{35}$$

or in terms of ω, $C = (\omega^2 - 1)^{1/2}/\omega$. Thus we see that the group velocity becomes imaginary for frequencies below the cutoff frequency $\omega = 1$, which will be seen later to have a significant effect on the solution. Further, it is noted that the maximum group velocity $C_M = 1$. In the previous problem the solution was exponentially small for $x > C_M t$, but here we shall find that $x = C_M t$ defines a distinct wavefront, ahead of which there is no disturbance. Consider $x > t$ in (34) and complete the path of integration L by the semicircle of radius $R \to \infty$ in the

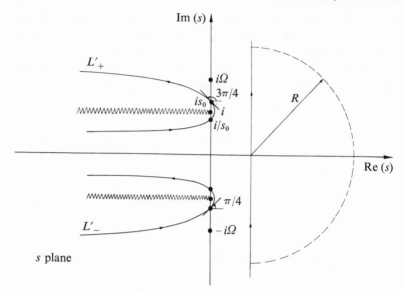

II.4. Path of integration for the Laplace transform representation of harmonic waves on a dispersive string.

right half-plane as shown in Figure II.4. On the semicircle, $(s^2 + 1)^{1/2} \to s$, and within the closed contour the integrand has no singularities. Jordan's lemma, therefore, may be applied; the result is $\phi = 0$ for $x > t$. The same result could be deduced directly from (2) by the method of characteristics (discussed in Chapter III).

To determine the disturbance immediately behind this wavefront, we let $t - x = \tau$ and take $\tau \to 0^+$. Then the major contribution to (34) may be found by expanding the integrand for large s and integrating term by term. Retaining terms up to order s^{-2} yields

$$\phi(x, \tau) \sim \frac{1}{2\pi i} \int_L \left(1 + \frac{i\Omega}{s}\right) \exp\left(s\tau - \frac{x}{2s}\right) \frac{ds}{s}, \tag{36}$$

which can be evaluated explicitly. Consider first

$$\Psi(x, t) = \frac{1}{2\pi i} \int_L \exp\left(s\tau - \frac{x}{2s}\right) \frac{ds}{s}$$

and let $s\tau - x/2s = (2\tau x)^{1/2}p$. Then

$$s = \left(\frac{x}{2\tau}\right)^{1/2} [p + (p^2 + 1)^{1/2}],$$

where we have chosen the root for s that gives $s = p$ as $|s| \to \infty$. The path of integration in the p plane can also be taken as a vertical line L lying to the right of the imaginary p axis. Thus

$$\Psi = \frac{1}{2\pi i} \int_L (p^2 + 1)^{-1/2} \exp [(2\tau x)^{1/2} p] \, dp = J_0[(2\tau x)^{1/2}],$$

since the Laplace transform of the Bessel function $J_0(t)$ is $(1 + s^2)^{-1/2}$. Now ϕ can be derived from (36) by the convolution theorem as

$$\phi \sim \Psi + i\Omega \int_0^\tau \Psi(x, \tau') \, d\tau'$$

$$= \Psi + i\Omega \left(\frac{2\tau}{x}\right)^{1/2} J_1[(2\tau x)^{1/2}]. \tag{37}$$

This expression is valid for all x in $0 \leq x$ and for $\tau \to 0^+$. If τx is very small, the Bessel functions can be approximated by the leading term in their power series expansions. This yields $\phi \approx 1 + i\Omega\tau$, which is identical to the leading terms in the expansion of the source function at $x = 0$. In other words, for fixed x, the wavefront transmits locally the initial form of the input signal. After a long time, however, such that the wavefront at $x = t$ has propagated a great distance and $\tau x >> 1$, the disturbance at the front will disperse into the usual harmonic oscillations.

The solution at arbitrary distances behind the wavefront can be determined for large x and t by the method of steepest descent. In (34), with $m = x/t < 1$ and

$$h(s) = \frac{s}{m} - (s^2 + 1)^{1/2}, \tag{38}$$

we find the cols are located at

$$s = \pm is_0 = \pm \frac{i}{(1 - m^2)^{1/2}}. \tag{39}$$

The two steepest descent paths through these cols (they both contribute disturbances of equal orders of magnitude) are shown in Figure II.4.

The upper path intersects the imaginary s axis at s_0 (>1) and $1/s_0$ (<1), while the slope of its tangent line at s_0 is $3\pi/4$. The lower path is the mirror image of the upper one. Note that for $1/s_0 < \Omega < s_0$, the original path L and the two paths $L_{\pm}'$, taken together, do not enclose the pole at $i\Omega$; but when $s_0 < \Omega$, the residue contribution at this pole must be included. Similarly, when $\Omega < 1/s_0$, a residue contribution is obtained, but it is exponentially small. Therefore, when we use (28), the imaginary part of the integral (34) becomes

$$\phi \sim H(\Omega - s_0) \sin \left[\Omega t - (\Omega^2 - 1)^{1/2} x\right] + H(s_0^{-1} - \Omega) \sin \Omega t \exp \left[-\right.$$

$$\left.(1 - \Omega^2)^{1/2} x\right] - \Omega\left(\frac{2}{\pi t}\right)^{1/2} \frac{m(1 - m^2)^{1/4}}{1 - \Omega^2 (1 - m^2)} \cos \left(\omega_0 t - k_0 x + \frac{\pi}{4}\right), \qquad (40)$$

where $H(y)$ is unity or zero for $y > 0$ or < 0, respectively. The case $\Omega = s_0$ is not covered by the solution above, since the pole and col at is_0 coincide, causing the last term in (40) to become unbounded. In (40) we also have $k_0 = m/(1 - m^2)^{1/2}$ and $\omega_0 = 1/(1 - m^2)^{1/2}$, where, as before, k_0 is the positive root of $C(k) = m$ and $\omega_0 = W(k_0)$. The group velocity for this problem was given in (35).

The first term in the asymptotic solution above represents a simple harmonic wavetrain of frequency Ω and wavenumber $k(\Omega)$ given by (35). Since it has a unit amplitude when $\Omega > s_0$, it then dominates the other terms and it is consequently called the "steady state signal" or simply the "signal," while the other terms comprise the transient response. The velocity of the front of the signal can be determined by setting $\Omega = s_0$ and using (39). We find

$$m = \frac{x}{t} = \frac{(\Omega^2 - 1)^{1/2}}{\Omega},$$

which is precisely the group velocity $\tilde{C}$, say, associated with the input frequency and is less than the maximum group velocity $C_M = 1$. Hence, the signal also propagates with its group velocity, even though its ripples are advancing with their phase velocity $\Omega/(\Omega^2 - 1)^{1/2}$, which exceeds $\tilde{C}$.

The second term in (40) occurs only when $\Omega < 1$, in which case the signal is "cutoff," since $s_0 > 1$. It is a standing wave decaying exponentially with distance from the origin, extending from $x=0$ to $x=(1-\Omega^2)^{1/2}t$, the latter position being determined from the relation $\Omega = 1/s_0$. Therefore, whether $\Omega > 1$ or $\Omega < 1$, either the first or second term, respectively, in (40) will correctly duplicate at $x=0$ the input signal, $\sin \Omega t$. This is because the input is assumed to be an unceasing source at $x=0$. However, the contrast between the type of response when $\Omega > 1$ or $\Omega < 1$ is striking.

The last term in (40) represents the familiar dispersed wave groups of order $t^{-1/2}$. For x and t large they provide the predominant disturbance if $\Omega < 1$; when $\Omega > 1$, they are subordinate in amplitude to the signal and will be observed merely as a transient forerunner, occupying the ever lengthening interval $(1 - \tilde{C})t$. Near $x=t$ this forerunner can be matched, in the usual manner, to the imaginary part of the more accurate "wavefront" solution (37).

To complete the asymptotic solution for $x > > 1$, we consider the case $\Omega \rightarrow s_0$. This will yield the behavior at the front of the steady state wavetrain. (The details of the analysis are quite lengthy and will be omitted. They involve a more accurate expansion of the integrand in (34) about the col, since it is essentially coincident with the pole at $i\Omega$.) The final result contains both a signal term and one proportional to a Fresnel integral. At the signal front ($\Omega = s_0$) these terms combine to form

$$\phi(x, t) \sim \tfrac{1}{2} \sin \left[\Omega t - (\Omega^2 - 1)^{1/2}x \right].$$

This is a group of oscillations having exactly one-half the amplitude of the fully developed steady state signal.

3e. Negative Group Velocity

We close this section with a few remarks about negative group velocity, which occurs when dW/dk becomes negative for positive values of k. In such a situation there would be no points of stationary phase in the first exponential in (20), but there could be at least one for the second exponential. Thus, with $m = x/t > 0$ and with $C(k_0) = -m$, the asymptotic form of the dispersive wavetrain becomes

$$\phi \sim \frac{\bar{f}(k_0)}{(2\pi t |C_0'|)^{1/2}} \cos \left(k_0 x + \omega_0 t \pm \frac{\pi}{4} \right)$$

instead of (21). The ripples in this group actually travel in the direction opposite to the group itself, but the group (and hence the energy it carries) is moving in the positive x direction, away from the source at the origin. The group's position is always given by $x = -C_0 t$ with C_0 negative.

In physical problems, negative group velocities have been discovered experimentally (Ibbetson and Phillips, 1967) and theoretically (Longuet-Higgins, 1968) in connection with waves in rotating fluids. Also, the dispersion curves for waves in elastic plates clearly indicate negative group velocities for many of the high-frequency modes.

Although the group and phase velocities are either in the same or the opposite directions for one-dimensional waves, the group velocity and phase velocity can occur in arbitrarily different directions for two- and three-dimensional waves in anisotropic media (Lighthill, 1960; Whitham, 1961). However, the group velocity vector still determines the direction of the energy transmission.

4. Kinematic Energy Approach for Dispersive Waves

It has now been shown that whatever the initial disturbance, or whatever the time variation of a source function, the main feature of the ensuing wave motion will be a periodic wavetrain. Furthermore, a group of ripples in the train having the same or essentially the same wavenumber will propagate with a group velocity corresponding to that wavenumber. Thus, we reiterate that k remains constant at positions given by $x = C(k)t$. Expressed mathematically,

$$dk = k_x \, dx + k_t \, dt = 0 \qquad \text{when} \ \frac{dx}{dt} = C,$$

or, in more compact notation,

$$k_t + C k_x = 0. \tag{41}$$

Equation (41) has another interesting interpretation. In an interval Δx there are $k\Delta x/2\pi$ ripples. Entering the interval are $\omega/2\pi$ ripples per unit time, and leaving the interval are $(\omega + \omega_x \Delta x)/2\pi$ ripples per unit time. Consequently, in the absence of a source (or sink) that could create (or destroy) ripples, there must be a balance equation for the rate of change of the number of ripples in the form

$$k_t + \omega_x = 0. \tag{42}$$

Since ω depends on x through its explicit dependence on k from the dispersion relation, it follows that $\omega_x = Ck_x$, and (41) and (42) are equivalent. Therefore, starting with a solution in the form of the wavetrain

$$\phi \sim a(x, t) \cos(kx - \omega t + \mu) \tag{43}$$

for large x and t, where μ is a phase constant, and applying the "conservation of ripples" (42), leads directly to the kinematic significance of group velocity. The value of k for given values of x and t is immediately found from $x = Ct$, and ω is then calculated from the dispersion relation.

The amplitude of the wavetrain (43) can be determined from a second conservation equation—the equation for the conservation of energy. Let us define the energy in an interval from x to $x + \Delta x$ to be

$$\Delta E = a^2 \Delta x$$

and suppose that waves with wavenumber k_0 are at $x = Ct$. Then at $x + \Delta x$ are waves with wavenumber $k_0 + \Delta k$, where $\Delta k = \Delta x / C_0' t$. Initially, the energy associated with the wavetrains having wavenumbers from k_0 to $k_0 + \Delta k$ can be shown to be

$$\Delta E = \frac{1}{2\pi} |\bar{f}(k_0)|^2 \Delta k,$$

where $\bar{f}(k)$ is the wavenumber spectrum (Fourier transform) of the initial waveform $f(x)$. Thus, if we use the result (derived in Chapter V) that the energy propagates with the group velocity, the above two expressions for ΔE must be equal, and we obtain

$$a = \frac{\bar{f}(k_0)}{(2\pi |C_0'| t)^{1/2}}. \tag{44}$$

Combining (41), (43) and (44), we produce the final kinematic energy result

$$\phi = \frac{\bar{f}(k_0)}{(2\pi |C'(k_0)| t)^{1/2}} \cos(k_0 x - \omega_0 t + \mu), \tag{43'}$$

where k_0 is the function of m determined by the equation $C(k_0) = x/t \equiv m$. Except for the fact that the phase μ is here arbitrary, this is identical to the stationary phase result. The main results for the dispersive wave motion may therefore be obtained with no appeal to the complicated asymptotic expansion of the solution in Fourier integral form.

On the other hand, since the starting point in the kinematic approach is the harmonic wavetrain (43), it cannot be used to extract the solution at caustics, or "wavefronts." The difficulty is that at a caustic, the harmonic wave response changes rapidly into a different form in a relatively short interval, exactly as through a boundary layer. The Airy waves, for example, effect a transition from a periodic motion to an exponential decaying precursor, while ahead of a distinct front the motion completely ceases.

It is still possible to avoid returning directly to the Fourier integral to determine the wavefront solution. The idea of a boundary-layer type solution for the governing equation of motion comes to mind, followed by a matching with the regular solution (43') to fix any unknown constants. S. Leibovich (private communication) has suggested the following perturbation approach to evaluate the Airy wave in the Love (or Boussinesq) equation problem that was treated in Subsection 3c, above.

The kinematic-energy theory would provide the previously derived solution (30) for $m = x/t < 1$, provided that the phase $\pi/4$ is replaced by the arbitrary phase μ. As $m \to 1$, this solution breaks down. (Note that the phase angle, $\mu = \pi/4$, would be unknown at this stage.) For $m \to 1$, or $x - t \to 0$, we must find a different solution that joins with the regions $m > 1$ and $m < 1$.

Our solution is to be valid for $x, t \to \infty$. We therefore introduce the small parameter ϵ as a measure of x^{-1}, t^{-1}, and the new time coordinate

$$T = \epsilon t.$$

At the caustic, $x - t$ is small compared to x and t separately (i.e., compared to ϵ^{-1}), and we formally recognize this fact by introducing the "boundary-layer" coordinate

$$X = \epsilon^\alpha(x - t).$$

Thus, in the transition region X is of order $(m-1)\epsilon^{\alpha-1}$ and α must be chosen so that this is $O(1)$, which implies that $0 < \alpha < 1$. With these preliminaries, (4) now becomes

$$\epsilon^{1-\alpha}\phi_{TT} - 2\phi_{XT} - \epsilon^{3\alpha-1}\phi_{XXXX} + 2\epsilon^{2\alpha}\phi_{XXXT} - \epsilon^{1+\alpha}\phi_{TTTT} = 0.$$

To obtain the detailed structure at the front of (30) we set $\epsilon = 0$ in the equation above, after first fixing α in $0 < \alpha < 1$ so that there will be at least one other term to balance $2\phi_{XT}$. The only possibility emerges as $\alpha = \frac{1}{3}$, and we have

$$\phi_{XT} = -\tfrac{1}{2}\phi_{XXXX}.$$

Also, ϕ must approach zero as $X \to \infty$, since there is no wavetrain ahead of $x = t$. Thus, by separation of variables and superposition, the preceding equation can be solved as

$$\phi = \frac{1}{2\pi}\int_{-\infty}^{\infty} A(p, \epsilon) \exp\left\{i\left[\left(\frac{2}{3}\right)^{1/3}Xp + \frac{T}{3}p^3\right]\right\} dp$$

$$= \frac{T^{-1/3}}{2\pi}\int_{-\infty}^{\infty} B(u, \epsilon) \cos\left[\left(\frac{2}{3T}\right)^{1/3}Xu + \frac{1}{3}u^3\right] du,$$

where $B(u, \epsilon) = A(u/T^{1/3}, \epsilon)$ is an unknown coefficient to be found by matching. We try $B(u, \epsilon) = B(\epsilon)$, so that

$$\phi = \frac{B}{T^{1/3}} Ai\left[\left(\frac{2}{3T}\right)^{1/3}X\right].$$

Replacing X by its definition in terms of the original variables, we obtain

$$\phi = \frac{B}{T^{1/3}} Ai\left[\left(\frac{2}{3T}\right)^{1/3}\frac{x-t}{\epsilon^{1/3}}\right].$$

Holding $x - t$ fixed and expanding this result as $\epsilon \to 0$ produces

$$\phi \sim \left(\frac{3}{2}\right)^{1/12}\frac{B}{(\pi t)^{1/2}\epsilon^{1/3}|\delta|^{1/4}}\cos\left[\left(2\frac{|\delta|}{3}\right)^{3/2}t - \frac{\pi}{4}\right]$$

for $x < t$, where $\delta = m - 1 = (x-t)/t$ as before. This should match the kinematic energy result (43'), which is given for this problem by

$$\frac{m^{-2/3}}{(6\pi t)^{1/2}(1 - m^{2/3})^{1/4}}\cos\left[t(1 - m^{2/3})^{3/2} - \mu\right]$$

as $m \to 1$. In terms of (X, T), $m = 1 + \epsilon^{2/3} X/T$. If we make this identification in the kinematic energy solution, expand for small ϵ and retain only the leading term, and finally revert back to x, t coordinates, we find that the resulting expression agrees with our expanded boundary layer result provided $\mu = \pi/4$, and $B = (\epsilon/12)^{1/3}$. This is a match in the sense of matched asymptotic expansions (cf. Van Dyke, 1964) and produces the missing phase angle in the kinematic energy solution; it also provides a valid transition through the caustic.

5. Effects of Dissipation

In every real physical system there is always a certain amount of energy dissipation. For the theory of dispersive waves in conservative systems to have useful applications, therefore, the results of this theory should closely approximate actual results when the energy losses are sufficiently small. For a linear system with a linear damping mechanism this is generally true. As an illustration, we consider the familiar case of viscous damping in which the frictional force is proportional to the material particle velocity. The linear Klein-Gordon equation (3) for waves on an elastically supported string becomes

$$\phi_{tt} - \phi_{xx} + \phi + 2\sigma\phi_t = 0 \tag{45}$$

when the string motion is damped by air friction. The parameter σ is a dimensionless damping coefficient. If we now substitute

$$\phi(x, t) = \exp(-\sigma t)\Psi(x, t)$$

into (45), there results

$$\Psi_{tt} - \Psi_{xx} + (1 - \sigma^2)\Psi = 0,$$

which is the regular Klein-Gordon equation for the string but with a spring constant that depends on σ. Clearly, for $\sigma < 1$ the solutions for Ψ are governed by the theory of dispersive waves, the group velocity being

$$C = \frac{k}{(k^2 + 1 - \sigma^2)^{1/2}}.$$

The complete solution for ϕ is seen to consist of exponentially damped wave groups. Note further that as $\sigma \to 0$, the group velocity becomes the same as that in the undamped system to order σ^2 and the complete solution becomes $\exp(-\sigma t)$ times the solution for the undamped system. On the other hand, for $\sigma > 1$ the group velocity is either imaginary or will exceed unity, which is the value of the actual wavefront velocity, as can be seen from (45). Hence the theory of dispersive waves does not apply in that case. When $\sigma = 1$, the interesting results are nondispersive waves, governed by the wave equation, with exponentially decaying magnitudes.

An alternate approach for defining a group velocity for (45) is to take

$$\phi(x, t) = \exp[i(kx - \omega t) - \alpha x],$$

which, when substituted into (45), yields

$$\alpha = \frac{\omega \sigma}{k} \quad \text{and} \quad \omega = (k^2 + 1 - \alpha^2)^{1/2}$$

or

$$\omega = k \left(\frac{k^2 + 1}{k^2 + \sigma^2} \right)^{1/2} \tag{46a}$$

The corresponding group velocity now becomes

$$C = \frac{k^4 + 2\sigma^2 k^2 + \sigma^2}{(k^2 + \sigma^2)^{3/2}(k^2 + 1)^{1/2}}, \tag{46b}$$

which is a real quantity. But, when $2k^2 > (5^{1/2} - 1)\sigma^2$, the group velocity above can again exceed the wavefront velocity of unity for suitable values of $\sigma > 1$.

Based on these calculations, it is certainly not clear what role, if any, a group velocity will play in the solution when $\sigma > 1$. We shall examine the matter for the special and convenient choice of $\sigma = \sqrt{2}$; and we shall use the Laplace transform method to solve (45) for a semi-infinite string $(0 < x)$ with

$$\phi(0, t) = \exp(i\Omega t)$$

—that is, for a sinusoidal input with frequency Ω. It is straightforward to show that the solution can be expressed by the inverse Laplace transform

$$\phi(x, t) = \frac{1}{2\pi i} \int_L \exp[st - (s^2 + 2\sigma s + 1)^{1/2} x] \frac{ds}{s - i\Omega}$$

or, with the translation $p = s + \sigma$ and with $\sigma^2 = 2$,

$$\phi(x, t) = \frac{1}{2\pi i} \exp(-\sigma t) \int_L \exp[pt - (p^2 - 1)^{1/2}x] \frac{dp}{p - (\sigma + i\Omega)}. \quad (47)$$

Figure II.5 shows the situation in the complex p plane. The cols of the exponent occur at

$$p = \pm p_0 = \pm(1 - m^2)^{-1/2}$$

for $m = x/t < 1$. (Note, for $m > 1$, $\phi = 0$, as can be shown by the same arguments used in Subsection 3d. Thus $x = t$ locates the position of the

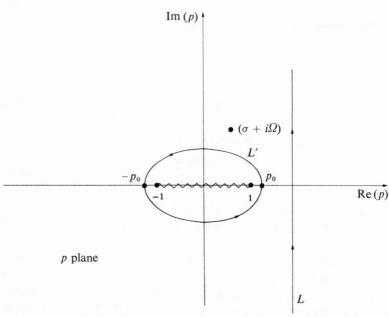

II.5. Path of integration for the Laplace transform representation of solutions to the dissipative Klein-Gordon equation (45).

wavefront.) The correct path of steepest descent passes through $p = p_0$ and is given by the ellipse

$$(1 - m^2)(m^2 u^2 + v^2) = m^2 \quad (48)$$

with $p = u + iv$.

If the ellipse encloses the pole at $\sigma + i\Omega$, then there is no residue contribution in deforming L to L'. The saddle point contribution obtained is exponentially small, being of the order $\exp\{-t[\sigma - (1 - m^2)^{1/2}]\}$. We can see this by taking the leading term in the expansion of the exponent in (47) about p_0. The response is the transient precursor to the main signal, but since it does not consist of periodic waves, there is indeed no group velocity for describing its motion.

On the other hand, when the pole lies outside the ellipse, we obtain the steady state signal that is given by the residue contribution

$$\phi(x, t) \sim \exp\{i[\Omega t - (\Omega^2 - 2i\sigma\Omega - 1)^{1/2}x]\}.$$

Thus, we may define the signal velocity to be the root obtained for m when the pole falls exactly on the ellipse—that is, when $u = \sigma$ and $v = \Omega$ is substituted into (48). The result (with $\sigma^2 = 2$) becomes

$$2m = [(\Omega^4 + 6\Omega^2 + 1)^{1/2} + 1 - \Omega^2]^{1/2},$$

which is less than the wavefront velocity for all values of Ω. Moreover, it does not correspond to the group velocity (46b) with $k = k(\Omega)$ from (46a).

As stated in Chapter I, Subsection 5a, the principal conclusion to be drawn here is that unless the dissipation is small, the utility of the concept of group velocity is lost.

CHAPTER III

Quasilinear Hyperbolic Systems that Result from Conservation Laws

Constantine M. Dafermos

The theory of a continuous medium rests upon a set of conservation laws and a set of constitutive equations. The former determines the nature of the theory (e.g., mechanical, thermodynamical, electromagnetic), while the latter characterizes the nature of the medium (e.g., elastic solid, viscous fluid). Conservation laws originally appear in integral form, since set rather than point functions have direct physical interpretation. In the class of smooth functions, however, the integral form is equivalent to a field equation. For constitutive assumptions of a particular type the system of field equations is hyperbolic.

A distinguishing feature of hyperbolic systems is the existence of special surfaces in space-time, which in physical applications manifest themselves as propagating waves. A study of the remaining chapters in this volume reveals that hyperbolic systems of field equations by no means monopolize wave phenomena in nature. In a broad class of problems not governed by hyperbolic field equations, but allowing wave phenomena, certain averaged quantities, such as "energy," are governed by hyperbolic equations. Consequently, hyperbolic systems are of fundamental importance in the study of wave propagation. See, for example, Chapter V.

The geometric features of solutions of hyperbolic systems have been studied for a long time, and the theory is reasonably complete. On the other hand the analytical investigation of the problem has met serious difficulties and is still in the developmental stage. The prime source of complication is the inherent property of nonlinear hyperbolic systems that, in general, the initial-value problem has no solution even if the initial data are very smooth. The notion of a solution has to be extended to include *generalized solutions*—discontinuous fields that satisfy the conservation law in its original integral form. As the class of solutions is extended, uniqueness is lost, and special criteria have to be applied to single out admissible solutions. In this connection the concepts of "entropy" and "irreversibility," which originated in thermodynamics, play a prominent role.

It is not the intent here to survey the general theory of hyperbolic systems, but rather to introduce the subject for students interested in applications to nonlinear wave propagation. Accordingly, only the aspects that have applications in research and continuum physics have been emphasized, and highly technical mathematical language has been avoided.

1. Conservation Laws and Field Equations

Let $\mathbf{x} = (x_1, \ldots, x_n)$ denote the generic point in "space" (for physical applications $n = 1$, 2, or 3) and t the "time" variable. A conservation law is an expression of the form*

$$\int_\Omega \mathbf{p}(\mathbf{x}, t) \Big|_{t_1}^{t_2} d\mathbf{x} + \int_{t_1}^{t_2} \oint_{\partial\Omega} \mathbf{A}(\mathbf{x}, t) \cdot \mathbf{v} \, dS \, dt + \int_{t_1}^{t_2} \int_\Omega \mathbf{g}(\mathbf{x}, t) \, d\mathbf{x} \, dt = \mathbf{0},$$

$$(1)$$

which holds for every time interval (t_1, t_2) and every domain Ω in space with smooth boundary $\partial\Omega$. Here $\mathbf{p}$, $\mathbf{g}$ are m-vector fields, $\mathbf{A}$ is an $m \times n$ matrix-valued field, and $\mathbf{v}$ is the unit normal on $\partial\Omega$. Equation (1) states that the change of the set function

$$\mathcal{P}(\Omega, t) = \int_\Omega \mathbf{p}(\mathbf{x}, t) \, d\mathbf{x}$$

between t_1 and t_2 is balanced by an influx through $\partial\Omega$ and a production in Ω during (t_1, t_2). In the conservation laws of continuum physics, typical components of $\mathcal{P}$ are mass, momentum, energy, electric charge, and so on.

The conservation laws are supplemented by constitutive assumptions, which characterize the nature of the continuous medium. The state of

---

* Here and throughout $d\mathbf{x}$ stands for $dx_1 \cdots dx_n$ and the notation $f(\mathbf{x}, t)\big|_{t_1}^{t_2}$ represents $f(\mathbf{x}, t_2) - f(\mathbf{x}, t_1)$.

the system is described by a state vector $\mathbf{u} = (u_1, \ldots, u_m)$ so that $\mathbf{p}$, $\mathbf{g}$, and $\mathbf{A}$ are determined by $\mathbf{u}$ through known differentiable *constitutive equations*,

$$\mathbf{p} = \hat{\mathbf{p}}(\mathbf{u}, \mathbf{x}, t), \quad \mathbf{g} = \hat{\mathbf{g}}(\mathbf{u}, \mathbf{x}, t), \quad \mathbf{A} = \hat{\mathbf{A}}(\mathbf{u}, \mathbf{x}, t). \tag{2}$$

The fundamental problem is to determine the state vector field $\mathbf{u}(\mathbf{x}, t)$ so that (1) is satisfied for every Ω, t_1, t_2 together with appropriate initial and boundary conditions.

For differentiable $\mathbf{u}(\mathbf{x}, t)$, (1) can be written in the form

$$\int_{t_1}^{t_2} \int_{\Omega} (\mathbf{p}_t + \operatorname{div} \mathbf{A} + \mathbf{g})\, d\mathbf{x}\, dt = 0,$$

where use has been made of the divergence theorem. The equation above will hold for arbitrary $\Omega \times (t_1, t_2)$ if, and only if,

$$\mathbf{p}_t + \operatorname{div} \mathbf{A} + \mathbf{g} = 0 \tag{3}$$

at every point $(\mathbf{x}, t)$. For a complete survey of conservation laws, constitutive equations, and field equations in various branches of continuum physics, we refer to Truesdell and Toupin (1960) and Truesdell and Noll (1965).

2. Hyperbolic Systems

Using the constitutive equations (2), one brings the field equation (3) into the form

$$\mathbf{B} \cdot \mathbf{u}_t + \sum_{j=1}^{n} \mathbf{C}^j \cdot \mathbf{u}_{x_j} + \mathbf{d} = 0 \tag{4}$$

where $\mathbf{B}$, $\mathbf{C}^1$, $\ldots$, $\mathbf{C}^n$ are $m \times m$ matrix-valued functions and $\mathbf{d}$ an m-vector field depending on $\mathbf{u}$, $\mathbf{x}$, t.

The quasilinear system (4) will be called *hyperbolic* if for every fixed $\mathbf{u}$, $\mathbf{x}$, t and any unit n-vector $\mathbf{\nu}$, the eigenvalue problem

$$\left(-\lambda \mathbf{B} + \sum_{j=1}^{n} \nu_j \mathbf{C}^j \right) \cdot \mathbf{v} = 0 \tag{5}$$

has m real eigenvalues, not necessarily distinct, and a set of m linearly independent eigenvectors. If, in particular, for every $\mathbf{u}$, $\mathbf{x}$, t, $\mathbf{\nu}$, the eigenvalues are distinct, the system is called *strictly hyperbolic*.

The next three sections discuss some fundamental properties of hyperbolic systems that provide the motivation for the above definition.

2a. Plane Waves

Consider the simplest case where $\mathbf{d} \equiv 0$ and the matrices $\mathbf{B}$, $\mathbf{C}^1$, ..., $\mathbf{C}^n$ are independent of $\mathbf{u}$, $\mathbf{x}$, t. For some unit vector $\mathbf{v}$ in space, let us seek a plane wave solution of (4) propagating in the direction of $\mathbf{v}$—that is,

$$\mathbf{u}(\mathbf{x}, t) = \mathbf{v} \exp\{ik(\mathbf{v} \cdot \mathbf{x} - \lambda t)\}, \tag{6}$$

with k, λ real numbers. One easily deduces that (6) is a solution of (4) if, and only if, λ and $\mathbf{v}$ satisfy (5). In other words, in the linear homogeneous case considered above, the system is hyperbolic if, and only if, in every direction in space m independent plane waves can propagate.

2b. Weak Waves

Let $\mathscr{S}$ be a smooth surface in space-time with equation $f(\mathbf{x}, t) = 0$. Suppose that $\mathbf{u}(\mathbf{x}, t)$ is a continuous m-vector field with the following properties: at every point not on $\mathscr{S}$, $\mathbf{u}$ is continuously differentiable, and (4) is satisfied; furthermore, one-sided limits of the space-time gradient of $\mathbf{u}$ exist on $\mathscr{S}$, but they are unequal so that first derivatives of $\mathbf{u}$ experience a finite jump across $\mathscr{S}$.

It is clear that $\mathscr{S}$ may be visualized as a one-parameter family of surfaces in space. Thus, in physical applications, $\mathscr{S}$ will appear as a propagating wave that will be designated weak so that it will be distinguished from stronger waves considered in Subsection 3b.

For each fixed t, the unit normal $\mathbf{v}$ on the wave [i.e., the space surface $f(\mathbf{x}, t) = 0$] has components

$$v_j = f_{x_j} / \left(\sum_{k=1}^{n} f_{x_k}^2 \right)^{1/2}, \tag{7}$$

with the appropriate interpretation if the denominator vanishes.

If V is the propagation speed of S in the direction of the spatial normal $\mathbf{v}$, then total derivative

$$\frac{df}{dt} = f_t + \sum_{j=1}^{n} \frac{dx_j}{dt} f_{x_j} = 0$$

for the trajectory $dx_j/dt = Vv_j$. Thus

$$V = -f_t / \left(\sum_{k=1}^{n} f_{x_k}^2 \right)^{1/2}. \tag{8}$$

Taking the jump of the lefthand side of (4) across $\mathscr{S}$ and using the fact that $\mathbf{B}$, $\mathbf{C}^1$, ..., $\mathbf{C}^n$ and $\mathbf{d}$ depend at most on $\mathbf{u}$, $\mathbf{x}$, t and are, therefore, continuous, one obtains

$$\mathbf{B} \cdot [\mathbf{u}_t] + \sum_{j=1}^{n} \mathbf{C}^j \cdot [\mathbf{u}_{x_j}] = 0, \tag{9}$$

where brackets denote the jump of the enclosed quantity across $\mathscr{S}$. Now $\mathbf{u}$ is continuous across $\mathscr{S}$, hence its tangential derivative is also. It follows that only the normal derivative of $\mathbf{u}$ can jump across $\mathscr{S}$. Since $(f_{x_1}, \ldots, f_{x_n}, f_t)$ is normal on $\mathscr{S}$, by (7) and (8), the vector $(\nu_1, \ldots, \nu_n, -V)$ is also normal on $\mathscr{S}$. Thus, if $\mathbf{v}$ denotes the jump of the normal derivative of $\mathbf{u}$ across $\mathscr{S}$,

$$[\mathbf{u}_{x_j}] = \nu_j \mathbf{v}, \quad [\mathbf{u}_t] = -V\mathbf{v}. \tag{10}$$

Substituting (10) into (9) yields

$$\left(-V\mathbf{B} + \sum_{j=1}^{n} \nu_j \mathbf{C}^j \right) \cdot \mathbf{v} = 0. \tag{11}$$

Comparing (11) with (5), we reach the following conclusion: the system (4) is hyperbolic if in every direction in space m independent weak waves can propagate. The eigenvalues of (5) give the speed of propagation, and the eigenvectors determine (up to a scaling constant) the amplitude of the waves.

An example of weak wave propagation will be presented in Subsection 3d.

2c. Characteristics

The mathematical formulation of a physical phenomenon in a continuous medium with field equations (4) usually leads to a *Cauchy problem*: a solution of (4) is sought with assigned values on a smooth surface $\mathscr{S}$ in space-time. It can be shown that the well-posedness of the Cauchy problem is intimately related to the following question: given that $\mathbf{u}(\mathbf{x}, t)$ is a solution of (4), does the restriction of $\mathbf{u}$ on $\mathscr{S}$—that is, the limitation of the argument of $\mathbf{u}$ to points on $\mathscr{S}$—determine the restriction of space-time gradient of $\mathbf{u}$ on $\mathscr{S}$? (See Chapter I, Subsection 3d, for examples of well-posed initial boundary value problems for the wave equation.)

Let $f(\mathbf{x}, t) = 0$ be the equation of $\mathscr{S}$ and $(y^1, \ldots, y^n, f)$ a curvilinear coordinate system that will have $\mathscr{S}$ as a coordinate surface. Applying the chain rule, (4) may be written in the new coordinate system as

$$(f_t \mathbf{B} + \sum_{j=1}^{n} f_{x_j} \mathbf{C}^j) \cdot \mathbf{u}_f + \sum_{\alpha=1}^{n} (y^{\alpha}_t \mathbf{B} + \sum_{j=1}^{n} y^{\alpha}_{x_j} \mathbf{C}^j) \cdot \mathbf{u}_{y^{\alpha}} + \mathbf{d} = 0. \qquad (12)$$

Now $\mathbf{u}_{y^{\alpha}}$ are tangential derivatives and hence are determined by the restriction of $\mathbf{u}$ on $\mathscr{S}$. Thus, the entire grad $\mathbf{u}$ will be determined if $\mathbf{u}_f$ can be computed from (12)—in other words, if the matrix

$$f_t \mathbf{B} + \sum_{j=1}^{n} f_{x_j} \mathbf{C}^j$$

is nonsingular. If on every point of $\mathscr{S}$ the matrix above is singular— that is, if f satisfies the differential equation

$$\det \{f_t \mathbf{B} + \sum_{j=1}^{n} f_{x_j} \mathbf{C}^j\} = 0 \qquad (13)$$

—$\mathscr{S}$ is called a *characteristic*.

A comparison of (13) with (5) reveals that the existence of characteristics depends crucially on the hyperbolicity of the system (4). Furthermore, comparing (13) with (11) and recalling (7) and (8), one deduces that every weak wave propagates along a characteristic. The eigenvalues of (5) are called *characteristic speeds* of (4).

In the case of one space variable ($n = 1$), every characteristic is the integral curve of the ordinary differential equation

$$\frac{dx}{dt} = -\frac{f_t}{f_x} = \lambda, \qquad (14)$$

where λ is a characteristic speed.

There is a rich geometric theory of characteristics with many applications; hopefully the examples given below will suggest its importance. A thorough study of concepts introduced in this section is contained in Courant and Hilbert (1962).

2d. Examples

The single equation in one space variable

$$u_t + \phi(u)_x = 0 \qquad (15)$$

arises in virtually every mathematical model of wave phenomena and may serve as a prototype of the entire theory. In this case, the eigenvalue problem (5) has the real eigenvalue $\lambda = \phi'(u)$, so that (15) is hyperbolic.

Equation (14) for the characteristics here takes the form

$$\frac{dx}{dt} = \phi'(u). \tag{16}$$

On the other hand, (15) yields $-u_t/u_x = \phi'(u)$, which shows that u is constant on characteristics. This, together with (16), implies that characteristics are straight lines.

As a second example, consider the system

$$
\begin{aligned}
v_t - \sigma(u)_x &= 0, \\
u_t - v_x &= 0,
\end{aligned}
\tag{17}
$$

which arises in fluid mechanics, nonlinear elasticity, and other branches of continuum mechanics. It is equivalent (with $v = w_t$, $u = w_x$) to the second-order wave equation

$$w_{tt} = \sigma(w_x)_x.$$

Here $\mathbf{B}$ is the identity matrix, and

$$\mathbf{C} = \begin{pmatrix} 0 & -\sigma'(u) \\ -1 & 0 \end{pmatrix},$$

with eigenvalues $\lambda_1 = +\sqrt{\sigma'(u)}$, $\lambda_2 = -\sqrt{\sigma'(u)}$ and corresponding eigenvectors $(-\sqrt{\sigma'(u)}, 1)$, $(+\sqrt{\sigma'(u)}, 1)$. Thus, the system is hyperbolic if $\sigma'(u) > 0$. The two families of characteristics will be the integral curves of the equations

$$\frac{dx}{dt} = \pm\sqrt{\sigma'(u)}. \tag{18}$$

Of course, it is no longer generally true that the characteristics are straight lines and that u or v remains constant on them. However, there is an analog of this last condition. In fact, let us define

$$r = v + \int_0^u \sqrt{\sigma'(\xi)}\, d\xi, \qquad s = v - \int_0^u \sqrt{\sigma'(\xi)}\, d\xi.$$

Using (17), we obtain

$$-\frac{r_t}{r_x} = \frac{v_t + \sqrt{\sigma'(u)}\, u_t}{v_x + \sqrt{\sigma'(u)}\, u_x} = -\sqrt{\sigma'(u)},$$

$$-\frac{s_t}{s_x} = \frac{v_t - \sqrt{\sigma'(u)}\, u_t}{v_x - \sqrt{\sigma'(u)}\, u_x} = \sqrt{\sigma'(u)},$$

which, on account of (18), implies that r is constant on every backward characteristic and s is constant on every forward characteristic. The functions r and s are called *Riemann invariants* of the system (17). For the general definition of Riemann invariants, see Lax (1957), and for their application to gas dynamics, see, for example, Courant and Friedrichs (1948).

3. Generalized Solutions

3a. *Nonexistence of Solutions*

Let us attempt to construct a solution of (15) with assigned initial conditions $u(x, 0) = u_0(x)$. As we have seen in the previous section, the characteristic issuing at any point x of the x axis is a straight line with slope $\phi'(u_0(x))$; hence it can be constructed in advance. On any point of this characteristic, u will be equal to $u_0(x)$. If it happens that for every pair of points (x_1, x_2),

$$(x_1 - x_2)\{\phi'(u_0(x_1)) - \phi'(u_0(x_2))\} \ge 0, \tag{19}$$

then the family of characteristics spans the entire half-plane $t \ge 0$ and generates a solution of the initial-value problem that is at least as smooth as $u_0(x)$.

Suppose, however, that (19) is violated for some pair of points (x_1, x_2). Then the characteristics issuing at x_1 and x_2 will collide at the point

$$\bar{x} = \frac{x_1\phi'(u_0(x_2)) - x_2\phi'(u_0(x_1))}{\phi'(u_0(x_2)) - \phi'(u_0(x_1))}, \quad \bar{t} = \frac{x_1 - x_2}{\phi'(u_0(x_2)) - \phi'(u_0(x_1))}.$$

If a solution of the initial-value problem exists, its value at $(\bar{x}, \bar{t})$ must be equal to both $u_0(x_1)$ and $u_0(x_2)$—which is, of course, impossible. Thus, if (19) is violated, no smooth global solution of the initial-value problem exists.

In the linear case, ϕ' is constant, so that the characteristics are parallel straight lines and a solution exists for arbitrary smooth initial data. If ϕ is nonlinear, however, it is only for a very limited class of initial data that (19) will not be violated. In other words, nonexistence of solutions of the initial-value problem for (15) is the rule rather than the exception. This is not a particular property of (15) but typical of general nonlinear hyperbolic systems. The nonexistence of a global solution of (17) is established in MacCamy and Mizel (1967).

3b. Solutions with Shocks and the Jump Conditions

The nonexistence of global solutions to an initial-value problem for the field equation (3) does not necessarily imply that the associated physical problem is not well posed. In fact, it is the conservation law (1) that has a direct physical interpretation, and clearly (1) may admit solutions in the class of discontinuous functions that are not solutions of (3). For historical reasons (the field equations were derived by alternative methods earlier than their counterparts in integral form), every solution of (1) is called a *generalized solution* of (3). For emphasis, every solution of (3) in the ordinary sense is called *classical*.

The class of solutions with weak waves considered in Section 2b provides an example of generalized solutions that are not classical. However, this class is not sufficiently broad to include all generalized solutions. We will consider here another class which, though not exhaustive, illuminates both the mathematical and the physical implications of the occurrence of generalized solutions.

Let $\mathscr{S}$ be a smooth surface in space-time and $\mathbf{u}$ a generalized solution of (3), which is continuously differentiable at every point not on $\mathscr{S}$. Furthermore, one-sided limits of $\mathbf{u}$ exist on $\mathscr{S}$ but they are unequal, so that $\mathbf{u}$ experiences a finite jump across $\mathscr{S}$. As in Section 2b, $\mathscr{S}$ can be interpreted as a propagating surface in space, which will be called a *shock wave*.

Repeating the procedure of Section 1 shows that the field equation (3) is satisfied at every point not on $\mathscr{S}$. Fix now some point $(\mathbf{x}, t)$ on $\mathscr{S}$. Let Ω be a small neighborhood of $\mathbf{x}$ and (t_1, t_2) a small neighborhood of t so that $\mathscr{S}$ divides the cylinder $\Omega \times (t_1, t_2)$ into two parts $\mathscr{V}_-$ and

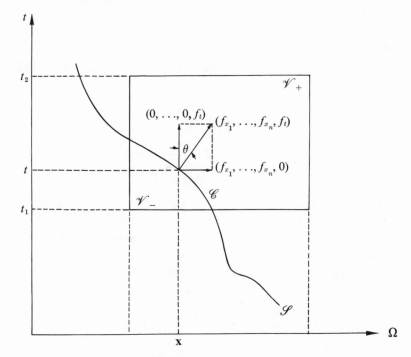

III.1. Shock wave in space-time.

$\mathscr{V}_+$ (Figure III.1). Let us integrate (3) over $\mathscr{V}_-$ and $\mathscr{V}_+$, separately, and then sum the two integrals. After an integration by parts,

$$\int_\Omega \mathbf{p}\, dx \,\Big|_{t_1}^{t_2} + \int_{t_1}^{t_2} \int_\Omega \mathbf{A} \cdot \mathbf{\nu}\, dS\, dt + \int_{t_1}^{t_2} \int_\Omega \mathbf{g}\, dx\, dt +$$

$$\oiint_\mathscr{C} ([\mathbf{p}] \cos\theta + [\mathbf{A}] \cdot \mathbf{\nu} \sin\theta)\, d\mathscr{S} = 0$$

where $\mathscr{C}$ is the part of $\mathscr{S}$ inside $\Omega \times (t_1, t_2)$ and θ is the angle between, the normal on $\mathscr{S}$ and the t axis (Figure III.1). As in Subsection 2b, brackets denote the jump of the enclosed quantity across $\mathscr{S}$ (with appropriate sign). Since $\mathbf{u}$ is a generalized solution of (3), the sum of the first three integrals vanishes. Moreover, the integrand of the integral over $\mathscr{C}$ is continuous on $\mathscr{S}$, so that, shrinking $\mathscr{C}$ around $(\mathbf{x}, t)$, one deduces

$$[\mathbf{p}] \cos\theta + [\mathbf{A}] \cdot \mathbf{\nu} \sin\theta = 0.$$

Let $f(\mathbf{x}, t) = 0$ be the equation of $\mathscr{S}$; then $(f_{x_1}, \ldots, f_{x_n}, f_t)$ is normal on $\mathscr{S}$, and $(0, \ldots, 0, f_t)$, $(f_{x_1}, \ldots, f_{x_n}, 0)$ are the orthogonal projections of this normal on the time axis and space, respectively (Figure III.1). Thus

$$\cot \theta = \frac{f_t}{\left(\sum_{k=1}^{n} f_{x_k}^{2}\right)^{1/2}} = -V,$$

where use has been made of (8). It follows that on every point of $\mathscr{S}$,

$$-V[\mathbf{p}] + [\mathbf{A}] \cdot \mathbf{\nu} = \mathbf{0}. \tag{20}$$

Conversely, it is easy to prove that if an m-vector field $\mathbf{u}$ is piecewise smooth, and if it satisfies (3) at every point of smoothness and (20) across any surface of discontinuity, then $\mathbf{u}$ is a generalized solution of (3). Equation (20) is called the Rankine-Hugoniot jump condition.

It should be noted that different conservation laws may lead to the same field equation. For example, both conservation laws

$$\int_{x_1}^{x_2} u \bigg|_{t_1}^{t_2} dx + \tfrac{1}{2} \int_{t_1}^{t_2} u^2 \bigg|_{x_1}^{x_2} dt = 0 \tag{21}$$

and

$$\int_{x_1}^{x_2} u^2 \bigg|_{t_1}^{t_2} dx + \tfrac{2}{3} \int_{t_1}^{t_2} u^3 \bigg|_{x_1}^{x_2} dt = 0 \tag{22}$$

lead to the same field equation

$$u_t + u u_x = 0. \tag{23}$$

However, the corresponding Rankine-Hugoniot jump conditions

$$-V[u] + \tfrac{1}{2}[u^2] = 0 \tag{24}$$

and

$$-V[u^2] + \tfrac{2}{3}[u^3] = 0 \tag{25}$$

are different. Thus, although the classes of smooth solutions of (21), (22) coincide, the classes of generalized solutions do not and (21), (22) are not equivalent. We will return to this point in discussing the concept of "entropy" (Subsection 4a).

A detailed study of singular surfaces in continuum physics can be found in Truesdell and Toupin (1960) and Truesdell and Noll (1965). For a systematic search of all conservation laws yielding an assigned field equation, see the survey article by Rozhdestvenskii (1960) and the references cited therein.

3c. Contact Discontinuities and Genuine Nonlinearity

In the linear case the constitutive equations (2) are

$$\mathbf{p} = \mathbf{B} \cdot \mathbf{u}, \qquad \mathbf{A} \cdot \mathbf{v} = \sum_{j=1}^{n} \nu_j \mathbf{C}^j \cdot \mathbf{u},$$

where the matrices $\mathbf{B}$, $\mathbf{C}^1$, ..., $\mathbf{C}^n$, introduced in Section 2, are independent of $\mathbf{u}$. Thus, the Rankine-Hugoniot jump condition (20) reads

$$\left(-V\mathbf{B} + \sum_{j=1}^{n} \nu_j \mathbf{C}^j \right) \cdot [\mathbf{u}] = \mathbf{0}.$$

Comparing this equation with (5), one reaches the conclusion that in linear hyperbolic systems, shocks propagate along characteristics.

This result does not carry over in general to nonlinear systems. However, it may occur that a shock wave is characteristic relative (at least) to the state on one side. A shock with this property is called a *contact discontinuity*. Since every shock in the linear case is a contact discontinuity, it is natural to call a system where no contact discontinuities are possible *genuinely nonlinear*.

The hyperbolic system (4) is, therefore, genuinely nonlinear if characteristics of the same family intersect; this must occur if

$$\mathbf{v} \cdot \operatorname{grad}_{\mathbf{u}} \lambda \neq 0$$

for every eigenvalue λ and associated eigenvector $\mathbf{v}$ of eigenvalue problem (5).

A simple computation shows that (15) and (17) are genuinely nonlinear if $\phi'' \neq 0$ and $\sigma'' \neq 0$, respectively.

The structure of solutions of genuinely nonlinear hyperbolic systems is considerably simpler, and for this reason the attention of analysts has been focused upon systems of this category. However, it should be emphasized that the class of genuinely nonlinear systems does not include many of the systems that arise in applications. For example, in the applications of (17) to nonlinear elasticity, the typical function σ (stress-strain curve) is neither convex nor concave. Similarly, the system of equations of gas dynamics, to be exhibited in Chapter IV, is not genuinely nonlinear.

For a discussion of contact discontinuities we refer to Lax (1957), who introduces the concept of genuine nonlinearity and studies its implications for the structure of solutions.

3d. Nonuniqueness of Generalized Solutions

By broadening the class of solutions of (3) to include generalized solutions, uniqueness of the solution to the initial-value problem is, in general, lost. This is demonstrated by the following simple example. Consider the conservation law (21), which leads to the field equation (23) and the Rankine-Hugoniot jump condition (24), under initial conditions

$$u(x, 0) = \begin{cases} 1, & x \le 0, \\ 0, & x > 0. \end{cases} \tag{26}$$

Condition (19) is violated for $x_1 \le 0 < x_2$ so that there is no smooth solution. However, the function

$$u(x, t) = \begin{cases} 1, & x \le \dfrac{t}{2}, \\ \\ 0, & x > \dfrac{t}{2}, \end{cases} \tag{27}$$

satisfies (23) for $x \neq t/2$ as well as (24) across the line $2x - t = 0$; hence it is a generalized solution of (23). (See Figure III.2.)

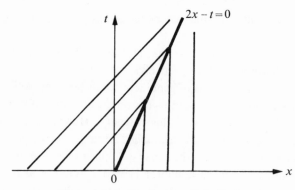

III.2. Admissible (or genuine) shock.

Consider now the same conservation law but with initial conditions

$$u(x, 0) = \begin{cases} 0, & x < 0, \\ 1, & x \geq 0, \end{cases} \tag{28}$$

which we obtain from (26) by reversing the direction of time. Thus, (27) immediately yields the following solution:

$$u(x, t) = \begin{cases} 0, & x < \dfrac{t}{2}, \\ 1, & x \geq \dfrac{t}{2}. \end{cases} \tag{29}$$

[See Figure III.3(a).]

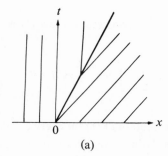 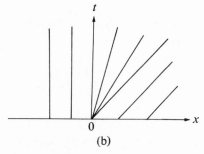

III.3. (a) Inadmissible shock.
(b) Centered simple wave.

Note, however, that the function

$$u(x, t) = \begin{cases} 0, & x < 0, \\ \dfrac{x}{t}, & 0 < x < t \\ 1, & x \geq t \end{cases} \tag{30}$$

is also a solution of (23), (28), which is continuous (except at $x = t = 0$). A solution of the form (30) in the sector $0 < x < t$ is called a *centered simple wave*. Note that the confining lines $x = 0$ and $x - t = 0$ of the centered simple wave are weak waves which, as expected, propagate along characteristics as depicted in Figure III.3(b).

Nonuniqueness has been established. Actually, combining (29) with (30) yields an infinite set of solutions of (23) and (28).

4. Admissibility Criteria for Generalized Solutions

A variety of criteria have been proposed to single out an admissible solution among the multiple, generalized solutions of the initial-value problem for a conservation law. Some of these criteria are mathematical in nature, relating to smoothness or stability properties of the solution; others reflect the requirement that admissible solutions should have an interpretation in the framework of the physical theory from which they have been derived. Ideally, the various admissibility criteria should be proved equivalent.

4a. The Entropy Criterion

An "entropy" condition associated with the conservation law (1) is an inequality

$$\int_\Omega \eta(\mathbf{x},\,t)\,\bigg|_{t_1}^{t_2}\,d\mathbf{x} + \int_{t_1}^{t_2} \oint_{\partial\Omega} \mathbf{Q}(\mathbf{x},\,t)\cdot\mathbf{\nu}\,dS\,dt + \int_{t_1}^{t_2} \int_\Omega \mathrm{h}(\mathbf{x},\,t)\,d\mathbf{x}\,dt \geq 0,$$
(31)

which must be satisfied by any admissible solution of (1) for every time interval $(t_1,\,t_2)$ and every smooth domain Ω in space. The *entropy* η, *entropy production* h, and *entropy flux* $\mathbf{Q}$ are determined by $\mathbf{u}$ through known differentiable constitutive equations

$$\eta = \hat{\eta}(\mathbf{u},\,\mathbf{x},\,t), \quad \mathrm{h} = \hat{\mathrm{h}}(\mathbf{u},\,\mathbf{x},\,t), \quad \mathbf{Q} = \hat{\mathbf{Q}}(\mathbf{u},\,\mathbf{x},\,t) \tag{32}$$

similar in nature to (2).

In the class of piecewise smooth fields $\mathbf{u}$ with shock waves, repeating the analysis in Sections 1 and 3b shows that (31) is equivalent to the field inequality

$$\eta_t + \mathrm{div}\,\mathbf{Q} + \mathrm{h} \geq 0, \tag{33}$$

which must hold at any point of smoothness, and the jump condition

$$-V[\eta] + [\mathbf{Q}]\cdot\mathbf{\nu} \geq 0 \tag{34}$$

across every shock.

Usually, the constitutive equations (32) are selected so that (33) is automatically satisfied whenever (3) holds. However, as we have seen in Subsection 3b, this does not imply that (34) is also satisfied. The entropy inequality for gas dynamics and general continuum thermo-dynamics has served as the prototype for developing the "entropy" criterion.

For (15), appropriate constitutive assumptions for an "entropy" condition are

$$\eta = - \int_0^u q(\xi)\, d\xi, \quad Q = - \int_0^u q(\xi)\phi'(\xi)\, d\xi, \quad h = 0, \tag{35}$$

where $q(\xi)$ is some strictly increasing function. With this choice, (33) reads $-q(u)u_t - q(u)\phi'(u)u_x \geq 0$ and hence is automatically satis-fied (as an equality) whenever (15) holds. On a shock, if u^- and u^+ are the limits of u from the left and the right, respectively, (34) yields

$$V \int_{u^-}^{u^+} q(\xi)\, d\xi - \int_{u^-}^{u^+} q(\xi)\phi'(\xi)\, d\xi \geq 0, \tag{36}$$

where $V = [\phi]\,[u]^{-1}$. It can be shown easily that if ϕ is strictly convex (strictly concave), (36) is equivalent to $u^- > u^+$ ($u^- < u^+$). Thus, solu-tion (27) of (23) is admissible, but (29) is not. For a detailed discus-sion of the entropy criterion, see Lax (1971) and the references cited there.

Unfortunately, for systems that are not genuinely nonlinear, the entropy criterion is not sufficiently powerful to rule out all inadmissible solutions. In order to remedy this difficulty, L. Leibovich and the author have proposed an alternative *entropy rate admissibility criterion*. Assume for normalization that $\hat{\eta}(0,\, \mathbf{x},\, t) = 0$. Consider a generalized solution $\mathbf{u}(\mathbf{x},\, t)$ of (3) that decays to zero sufficiently fast as $|\mathbf{x}| \to \infty$. Then the total entropy

$$H_{\mathbf{u}}(t) = \int_{E^n} \hat{\eta}(\mathbf{u}(\mathbf{x},\, t),\, \mathbf{x},\, t)\mathrm{d}\mathbf{x}$$

is finite and, by (31), increasing with t. Call $\mathbf{u}(\mathbf{x},\, t)$ admissible if at every $\tau \geq 0$ the rate of entropy increase $D^+H_{\mathbf{u}}(\tau)$—that is, the derivative of $H_{\mathbf{u}}(t)$ on the right at $t = \tau$—is greater than or equal to $D^+H_{\hat{\mathbf{u}}}(\tau)$ for any other generalized solution $\hat{\mathbf{u}}(\mathbf{x}, t)$ of (3) that coincides with $\mathbf{u}(\mathbf{x}, t)$ on $\Omega \times [0, \tau]$. Roughly speaking, we require that for an admissible solution the total entropy must increase as fast as possible. It can be shown that with the choice (35) for η, this criterion singles out a unique admissible solution of the initial-value problem for (15) even if ϕ is not convex or concave. The criterion also has been used successfully for (17), thus strongly indicating it can be applied to more general systems.

The entropy inequality in gas dynamics and general continuum mechanics is discussed by Courant and Friedrichs (1948) and Truesdell and Toupin (1960).

4b. The Smoothness Criterion

According to this criterion, every classical solution is admissible. A discontinuous generalized solution is admissible only if it can be approximated in an appropriate sense by a sequence of classical solutions.

4c. The Shock Admissibility Criterion

This criterion is applicable only to genuinely nonlinear, strictly hyperbolic systems (though recently modifications applicable to a somewhat broader class of systems have been proposed). A generalized solution with shocks is designated admissible if every shock propagating, say, in the direction $\mathbf{v}$ with speed V and joining the states $\mathbf{u}^-$ and $\mathbf{u}^+$ satisfies the following condition: there is an index k between 1 and n such that

$$\lambda_{k-1}(\mathbf{u}^-) < V < \lambda_k(\mathbf{u}^-),$$

$$\lambda_k(\mathbf{u}^+) < V < \lambda_{k+1}(\mathbf{u}^+),$$

where $\lambda_1 < \lambda_2 < \ldots < \lambda_n$ are the characteristic speeds of (4). In particular, for (15) with ϕ strictly convex (strictly concave), a shock is admissible if $u^- > u^+$ ($u^- < u^+$), which is also in agreement with the entropy condition. The shock admissibility criterion was introduced and analyzed by Lax (1957). The motivation is furnished by the theory of free boundary problems, and its equivalence with the entropy criterion has been established in general by Lax (1971).

*4d. The Criterion of Proper Embedding of the Physical Model in a More
Complete One*

Chapter IV contains a discussion of a piston problem for a gas and
demonstrates that under certain assumptions the (scaled) speed u of the
gas satisfies the parabolic equation

$$u_t + uu_x = \frac{\delta}{2} u_{xx} , \qquad (37)$$

where δ is proportional to the longitudinal viscosity and is positive and
usually very small. If the gas is inviscid, (37) reduces to (23). It can be
shown that the initial-value problem for (37) always has a unique
solution that is smooth. In order that the theory for an inviscid gas be
properly embedded in the theory for a viscous gas, the following condition must hold: if u is an admissible solution of (23) and u_δ the solution
of (37) with the same initial conditions, then $u_\delta \to u$ as $\delta \to 0$ (in an
appropriate sense). The limiting behavior of solutions of (37) as $\delta \to 0$
was studied by Hopf (1950). It has been proved that a solution of (23)
is admissible according to this criterion if it is admissible according to
the entropy criterion [for example, see Lax (1971)].

In the example above, the continuum theory was embedded in a more
complete continuum theory. In other problems the continuum theory is
visualized as a limiting case of a discretized theory governed by difference or difference-differential equations. Then the admissible solution
of the hyperbolic system must be the limit of solutions of a sequence of
systems of difference or difference-differential equations.

For a detailed discussion of the material in this subsection, see
Chapter IV. An extension to systems of two equations is studied by
Conley and Smoller (1970).

5. The Riemann Problem

Consider a homogeneous hyperbolic system in one space variable:

$$\mathbf{u}_t + \mathbf{C}(\mathbf{u}) \cdot \mathbf{u}_x = 0 \qquad (38)$$

with initial conditions

$$\mathbf{u}(x, 0) = \begin{cases} \mathbf{u}^-, & x \le 0 \\ \mathbf{u}^+, & x > 0, \end{cases} \qquad (39)$$

where $\mathbf{u}^-$ and $\mathbf{u}^+$ are constant vectors.

The study of this initial-value problem, which is called a Riemann problem, illuminates the concept of genuine nonlinearity. Moreover, its solution serves as a building block for constructing solutions of (38) under more general initial conditions. Obviously, if $\mathbf{u}(x, t)$ is a (generalized) solution of (38), (39), and α is any positive constant, the function $\mathbf{u}(\alpha x, \alpha t)$ is also a solution. Assuming that there is a unique admissible solution and that admissibility is preserved under a uniform expansion of coordinates (without reflections), it follows that $\mathbf{u}(x, t) = \mathbf{u}(\alpha x, \alpha t)$—in other words, $\mathbf{u}(x, t)$ is a function of the single variable x/t.

The solution of a Riemann problem for (23) is given in Subsection 3d. The admissible solution consists of two constant states joined either by a shock [in the case of initial conditions (26)] or by a centered simple wave [in the case of initial conditions (28)]. It is easy to see that the same type of solution exists for the Riemann problem for the more general (15) with ϕ strictly convex or strictly concave.

The concept of a centered simple wave can be extended to general systems of the form (38), and the following generalization of the result above has been established: if the system (38) of n equations is genuinely nonlinear and $\mathbf{u}^+$ is sufficiently close to $\mathbf{u}^-$, there is a solution of the Riemann problem that consists of $n + 1$ constant states joined by simple waves or admissible shocks (Lax, 1957). For systems of two equations, the restriction that $\mathbf{u}^+$ is close to $\mathbf{u}^-$ is not necessary (Smoller, 1969).

For systems that are not genuinely nonlinear, the problem is more complicated, and only partial results are available. For (15) without convexity assumptions on ϕ the Riemann problem has been solved, and the solution again consists of two constant states—now, however, joined by a shock fan (a sequence of adjacent shocks and simple waves). Similar solutions of the Riemann problem for (17) without any convexity assumptions (Leibovich, 1973) as well as for the system of equations of gas dynamics (Wendroff, 1972), have recently been established.

6. Existence, Uniqueness, and Computation of Generalized Solutions

For the single equation

$$u_t + \sum_{j=1}^{n} \phi_j(u, \mathbf{x}, t)_{x_j} + g(u, \mathbf{x}, t) = 0 \tag{40}$$

the theory of existence and uniqueness is now available in definitive form. For initial data in the class of bounded measurable functions, there is a unique admissible solution $u(x, t)$ such that for each fixed t, $u(\cdot, t)$ is a bounded measurable function on E^n that depends continuously on the initial data (in an appropriate sense). Furthermore, if the initial data are of bounded variation in the sense of Tonelli-Cesari, then $u(\cdot, t)$ is also of bounded variation in that sense. From the point of view of applications to continuum physics the class of functions of bounded variation is more appropriate, since it is the natural class in which shocks can be defined and studied.

The results above have been established independently by the method of vanishing viscosity (motivated by the discussion in Subsection 4d) (Volpert, 1967; Kružkov, 1970), the method of regularization (related to the smoothness criterion, Subsection 4b) (Kuznetsov, 1967), and the method of finite differences (Conway and Smoller, 1966), which also provides a scheme for the numerical construction of the solution.

Equation (15), the one-dimensional form of (40), has been studied for a long time. Several converging difference schemes have been proposed, (e.g., Oleinik, 1957). Solutions can also be constructed by the method of the potential (Rozhdestvenskii, 1960) and the method of polygonal approximations (Dafermos, 1972). If ϕ is strictly convex or strictly concave, an explicit solution is available (Lax, 1957).

In contrast, the existence theory for systems of more than one equation is relatively underdeveloped. The most general known theorem establishes the existence of a generalized solution to a genuinely nonlinear system of the form (38) under initial conditions with sufficiently small oscillation and total variation (Glimm, 1965). For systems of two equations, the assumptions have been relaxed so that the total variation of the initial data need not be small or even bounded (Glimm and Lax, 1970). The existence of solutions for genuinely nonlinear (and slightly more general) systems has also been established for initial conditions that do not have small oscillation but satisfy, instead, rather stringent monotonicity conditions (Zhan Tong and Guo Yu-Fa, 1965; Johnson and Smoller, 1967, 1969). The interaction of various combinations of waves has been studied in the same framework (Greenberg, 1970).

To this author's knowledge, the only available existence theorem for systems under general initial conditions pertains to (17) in the special case $\sigma(u) = -a/u$ (Nishida, 1968). For systems that are not genuinely nonlinear, no existence theorem has yet been established, although a great number of particular solutions are available—namely, for the equations of gas dynamics.

Uniqueness theorems in the class of piecewise smooth functions are known for fairly general systems (Rozhdestvenskii, 1960). However, this class of functions is not sufficiently broad to include all generalized solutions (for example, it excludes solutions with centered simple waves). The only uniqueness theorem in an appropriately broad function class has been established for the system (17) (as well as its extension to several space variables) under the assumption that σ is convex or concave (Oleinik, 1957a; Hurd, 1969, 1970).

CHAPTER IV

Examples of Dissipative and Dispersive Systems Leading to the Burgers and the Korteweg-deVries Equations

Sidney Leibovich and A. Richard Seebass

One of the most characteristic features of wave motions is their persistence. Waves typically survive for long periods and can transmit disturbances over very long distances. In fact, waves assume their most distinctive forms after traveling a "large" distance from the region in which they are generated. An important and difficult mathematical question asks for the large-time behavior of small-amplitude waves experiencing weak dissipation. These restrictions are not as special as they might seem at first glance. Granted the observed persistence of waves, the large-time and weak-dissipation restrictions are quite natural. The small-amplitude condition allows one to begin with a linear mathematical problem. Fortunately, the assumption of small amplitude usually makes physical sense, for if one waits long enough, all unforced waves will have small amplitudes and all forced waves that propagate in more than one dimension are attenuated by geometrical spreading.

In this chapter we develop the mathematical techniques for dealing with typical physical systems within the framework of the remarks above and apply them to two examples. The first, from gas dynamics, has weak dissipation measured by the dimensionless parameter Re^{-1}; the second, from shallow-water waves, is conservative but has weakly dispersive tendencies measured by a parameter δ. In both cases an amplitude measure ϵ is small but nonzero. In the cases of greatest interest, as we shall see, nonlinear equations must be faced for times of order ϵ^{-1}. Both equations can be summed up in the composite model equation

$$u_t + \epsilon u u_x = \mathrm{Re}^{-1} u_{xx} + \delta u_{xxx}.$$

Burgers equation obtains for $\delta = 0$, $\epsilon > 0$, $Re^{-1} > 0$, while the Korteweg-deVries equation obtains for $Re^{-1} = 0$, $\epsilon \neq 0$, $\delta \neq 0$. In either event, non-linear effects are important only if the initial disturbance is large enough to make ϵ comparable to Re^{-1} or δ. (The "initial" time can be any instant at which one wishes to begin calculating the subsequent behavior.) Although the two equations are derived in the context of special cases, it should be emphasized that they arise in a wide variety of different physical contexts (see Chapter VIII for a citation of other examples) and so they are, in a sense, typical of nonlinear dissipative and dispersive systems.

1. Perturbation Techniques

The "method of multiple scales" that we use to develop the equations is expected to be asymptotically correct for large times as ϵ, Re^{-1}, $\delta \rightarrow 0$. This technique is in wide use but, like many asymptotic theories, it does not have a rigorous mathematical foundation. The assumption is, "If it works, it's right." The method of multiple scales is used here to systematically reduce a system of equations depending upon several small parameters to an equation for a single unknown function. The technique we use is a modification of the procedure of Taniuti and Wei (1968) and had its origins in the work of Kevorkian (1966), Cole (1968) and Lick (1969); see also Lick (1970).

1a. Two-Timing

Before introducing our model system of equations, we may briefly illustrate the need for multiple time-scaling by a simple example involving a single equation

$$u_t + Lu = \epsilon Mu, \tag{1}$$

where L is a linear operator involving only operations with respect to the coordinate x, M is some other, possibly nonlinear, operator and ϵ is a small parameter.

For simplicity, consider the initial-value problem defined on the real line $-\infty < x < \infty$, with $u(x, 0) = f(x)$ and such that the Fourier transform

$$\tilde{f}(k) = \int_{-\infty}^{\infty} f(x)e^{-ikx}\,dx$$

exists. We will assume that the Fourier transform of Lu is $i\omega(k)\bar{u}(t; k)$, where $\bar{u}$ is the transform of $u(x, t)$. This will be the case, for example, when L is a differential operator or an integral operator of the convolution type.

Attempting to solve (1) by a regular perturbation procedure, we put

$$u(x, t; \epsilon) = \sum_{n=0}^{\infty} \epsilon^n u_n(x, t), \qquad (2a)$$

which has the Fourier transform

$$\bar{u}(t; k; \epsilon) = \sum_{n=0}^{\infty} \epsilon^n \bar{u}_n(t; k). \qquad (2b)$$

The transform of (1) is thus

$$\bar{u}_t + i\omega(k)\bar{u} = \epsilon \overline{Mu}. \qquad (3)$$

Substituting (2b) into (3) and equating coefficients of power of ϵ, we find that the first two equations of the infinite set produced are

$$u_{0_t} + i\omega(k)\bar{u}_0 = 0 \qquad (4)$$

and

$$\bar{u}_{1_t} + i\omega(k)\bar{u}_1 = \overline{Mu_0}. \qquad (5)$$

We assume that

$$\overline{Mu_0} = im(k)\bar{u}_0, \qquad (6)$$

which will occur, for example, for linear operators in the same class as L or (with a generalized definition of the Fourier transform) under the conditions of "nonlinear resonance" discussed in Chapter VII.

The solution to (4) satisfying the initial condition is

$$\bar{u}_0 = \bar{f}(k)e^{-i\omega(k)t}.$$

Substituting $\bar{u}_0$ into (5), we have

$$u_{1_t} + i\omega(k)\bar{u}_1 = i\bar{f}m(k)e^{-i\omega(k)t},$$

which has the solution

$$\bar{u}_1 = i\bar{f}m(k)te^{-i\omega(k)t}, \qquad (7)$$

where the initial condition $\bar{u}_1(0; k) = 0$ has been applied.

The first two terms of (2a) are thus

$$\bar{u} = \bar{f}(k)e^{-i\omega(k)t} [1 + i\epsilon t m(k) + \cdots], \qquad (8)$$

and consequently the error of the second term relative to the first is of $O(\epsilon t)$. For a wide class of functions $m(k)$, this implies the same relative error in (2a). Frequently, one can expect such errors to occur in higher-order terms of the series as well.

For fixed ϵ, the product ϵt is no longer small when $t = O(\epsilon^{-1})$, and so the series (2) is no longer valid. The form of the solution (8) suggests that u depends upon both t, which we now call the *fast time*, and ϵt, which we denote by τ and call the *slow time*. The essence of the method of multiple scales is the formal introduction of this notion with u assumed to be a function of three *independent* variables x, t, and $\tau = \epsilon t$, that is, x and two time variables:

$$u(x, t; \epsilon) = U(x, t, \tau; \epsilon).$$

The method "works" when a regular perturbation series in ϵ for U, valid to times of order ϵ^{-1}, may be found by appropriate choices of the τ dependence of U. Thus it is assumed that $U(x, t, \tau; \epsilon)$ has a series representation of the form (2a) with the Fourier transform

$$\overline{U}(t, \tau; k; \epsilon) = \sum_{n=0}^{\infty} \epsilon^n \overline{U}_n(t, \tau; k). \tag{9}$$

Notice that time derivatives have the form

$$\bar{u}_t = \overline{U}_t + \epsilon \overline{U}_\tau. \tag{10}$$

Returning to (3), substituting the series for $\overline{U}(t, \tau; k; \epsilon)$ and noting (10), we find that the equations for the first two coefficients are

$$\overline{U}_{0_t} + i\omega(k)\overline{U}_0 = 0$$

and

$$\overline{U}_{1_t} + i\omega(k)\overline{U}_1 = im(k)\overline{U}_0 - \overline{U}_{0_\tau}.$$

The solution for $\overline{U}_0$ satisfying the initial condition is now

$$\overline{U}_0 = \bar{f}(k)A_0(\tau; k)e^{-i\omega(k)t},$$

which is the same as before except for the coefficient $A_0(\tau; k)$. Aside from the restriction $A_0(0; k) = 1$ required by the initial condition, this coefficient is an arbitrary function of τ at this point. Exploitation of this arbitrariness allows the solution to be made uniformly valid to $t = O(\epsilon^{-1})$.

The righthand side of the equation for $\overline{U}_1$, the cause of the previous difficulty, is now

$$\overline{f}(k)e^{-i\omega(k)t}[im(k)A_0(\tau; k) - A_{0_\tau}].$$

This will vanish if

$$A_0 = b(k)\, e^{i\tau m(k)},$$

and we must choose $b(k) = 1$ in order to satisfy initial conditions. The series (9) then becomes

$$\overline{U} = \overline{f}(k)\exp\{-i[t\omega(k) - \tau m(k)]\} + \epsilon A_1(\tau; k)e^{-it\omega(k)} + \cdots. \qquad (11)$$

Frequently this procedure can be continued indefinitely, and if so, the series (9) is valid for all time. In carrying out the procedure, we will determine $A_n(\tau; k)$ by considering the equations for $\overline{U}_{n+1}$ and so forth.

By replacing τ by ϵt and expanding for ϵt small, it may be seen that the first term of (11) reproduces the first two terms of the series (8), the important distinction between the two being that (11) is valid for $t = 0(\epsilon^{-1})$ while (8) is not.

1b. Reduction of Systems of Equations

We now consider those systems of nonlinear equations depending upon one or more small parameters that permit nondispersive, non-dissipative wave motions when all of the small parameters vanish. As in the preceding subsection, t, time, and the single spatial coordinate x will be the independent variables. It will be assumed that the system may be put in the form

$$\mathbf{u}_t + \mathbf{C}(\mathbf{u}) \cdot \mathbf{u}_x = \sum_{k=1}^{m} \mu_k \mathfrak{D}_k(\mathbf{u}, x, t), \qquad (12)$$

where $\mathbf{u}$ is the column vector

$$\mathbf{u} = \begin{pmatrix} u_1 \\ \vdots \\ u_n \end{pmatrix},$$

$\mathbf{C}$ is an $N \times N$ matrix that may depend on $\mathbf{u}$, the μ_k are the small parameters, and the $\mathfrak{D}_k$ are vector-valued operators on $\mathbf{u}$ that depend on $\mathbf{u}$, x and t; the operation on $\mathbf{u}$ is not shown explicitly. For example, $\mathfrak{D}_k$ could be $\mathbf{I} \cdot (\partial^2 \mathbf{u}/\partial x^2)$ where $\mathbf{I}$ is the identity matrix.

The motion is assumed to be a small disturbance, measured by the parameter ϵ, to a constant state $\mathbf{u}_c$ ($\mathbf{u}_c$ is therefore assumed to be a solution, so that $\mathfrak{D}_k(\mathbf{u}_c, x, t) = 0$); an expansion about $\mathbf{u}_c$ in the small parameters is attempted in the form

$$\mathbf{u} = \mathbf{u}_c + \epsilon\{\mathbf{u}_0 + \epsilon\mathbf{u}_1 + \cdots + \sum_{j=1}^{m} \mu_j \mathbf{u}_{j+1} + \cdots\}, \tag{13}$$

where omitted terms are proportional to quadratic and higher powers of ϵ and μ_k. (Owing to calculational complexity, only the terms explicitly shown are usually computed.) Substituting (13) into (12) and neglecting the higher-order terms, we find

$$\mathbf{u}_{0_t} + \mathbf{C}(\mathbf{u}_c) \cdot \mathbf{u}_{0_x} = 0. \tag{14}$$

In order that (12) allow wave motion as $\epsilon \to 0$, we assume that the matrix $\mathbf{C}$ has the expansion

$$\mathbf{C}(\mathbf{u}_c + \epsilon\{\mathbf{u}_0 + \cdots\}) = \mathbf{C}_0 + \epsilon\mathbf{C}_1(\mathbf{u}_0, \mathbf{u}_c) + \cdots,$$

where $\mathbf{C}_0 = \mathbf{C}(\mathbf{u}_c)$ is a matrix with constant coefficients that has N real and distinct eigenvalues $\lambda_1, \ldots, \lambda_N$; in other words,

$$\mathbf{C}_0 \cdot \mathbf{\imath}_i = \lambda_i \mathbf{\imath}_i, \tag{15}$$

where the $\mathbf{\imath}_i$ are the right (column) eigenvectors corresponding to the λ_i. The solution to (14) is

$$\mathbf{u}_0 = \mathbf{\imath}_i U(x - \lambda_i t), \tag{16}$$

where $\mathbf{\imath}_i$ and λ_i satisfy (15); this is verified by substituting (16) into (14), which yields

$$[-\lambda_i \mathbf{\imath}_i + \mathbf{C}_0 \cdot \mathbf{\imath}_i] U_{X_i} = 0,$$

where $X_i = x - \lambda_i t$. In view of (15), (16) is a solution for arbitrary functions $U(X_i)$ and for $N = 2$ is the familiar elementary solution of the linearized wave equation. To proceed with higher-order corrections we focus upon one of the N solutions (16). In what follows, it will be assumed that one has been selected, and we shall omit the subscripts on the selected pair $(\lambda, \mathbf{\imath})$. In dealing with the N eigenvalues one at a time, we must confine ourselves to the asymptotic behavior of a single mode for large time or to initial data close to the traveling wave (16). In the latter case, the initial data should, to lowest order, depend only on

$x - \lambda t$, and most applications of multiple scale methods to partial differential equations have the same restriction. The general initial-value problem requires all N eigenvalues for its solution. For an approach leading to a solution uniformly valid in time for arbitrary initial data, see Chikwendu and Kevorkian (1972). Boundary conditions may also require the use of more than one mode. Lesser and Seebass (1968) and Hung and Seebass (1973) treated two such problems by matched asymptotic expansions; the multiple scale technique would have been a more natural one for these problems.

Equations for the higher-order terms are found by substituting the series (13) into (12) and equating coefficients of ϵ, μ_1, ..., μ_m. With $\mathfrak{D}_k(\mathbf{u}_c, x, t) = 0$ we assume that $\mathfrak{D}_k$ has the property that $\mathfrak{D}_k(\mathbf{u}_c + \epsilon\mathbf{u}_0, x, t) = \epsilon\mathbf{D}_k(\mathbf{u}_c, \mathbf{u}_0, x, t) + \cdots$. With this understood, the equations are

$$\mathbf{u}_{1_t} + \mathbf{C}_0 \cdot \mathbf{u}_{1_x} = -\mathbf{C}_1(\mathbf{u}_0, \mathbf{u}_c) \cdot \mathbf{u}_{0_x} \tag{17a}$$

and

$$\mathbf{u}_{j_t} + \mathbf{C}_0 \cdot \mathbf{u}_{j_x} = \mathbf{D}_{j-1}(\mathbf{u}_c, \mathbf{u}_0, x, t), \qquad j = 2, \ldots, m+1. \tag{17b}$$

Each of the vector equations (17) presumably represents a solvable problem for the N associated functions. Without generalizing (16) to account for multiple time-scaling, however, the solutions are not valid unless $t \ll \mu^{-1}$, where $\mu = \max\{\epsilon, \mu_1, \ldots, \mu_m\}$. To clarify this point, consider the equation for $\mathbf{u}_1$. Since $\mathbf{u}_0 = \imath U(x - \lambda t)$, it is convenient to transform coordinates to

$$X = x - \lambda t, \tag{18}$$

which changes (17a) to

$$\mathbf{u}_{1_t} + (\mathbf{C}_0 - \lambda\mathbf{I}) \cdot \mathbf{u}_{1_X} = -\mathbf{C}_1(\mathbf{u}_0, \mathbf{u}_c) \cdot \mathbf{u}_{0_X},$$

where $\mathbf{I}$ is the identity matrix. If this equation is premultiplied by the left eigenvector $\boldsymbol{\ell}$ corresponding to the eigenvalue λ—that is, the vector satisfying the equation

$$\boldsymbol{\ell} \cdot (\mathbf{C}_0 - \lambda\mathbf{I}) = 0,$$

the resulting equation is

$$\boldsymbol{\ell} \cdot \mathbf{u}_{1_t} = -\boldsymbol{\ell} \cdot \mathbf{C}_1 \cdot \mathbf{u}_{0_X}.$$

In general, the quantity $\boldsymbol{\ell} \cdot \mathbf{C}_1 \cdot \mathbf{u}_{0X} \neq 0$, and since the righthand side is a function of X alone,

$$\boldsymbol{\ell} \cdot \mathbf{u}_1 = -t\,\boldsymbol{\ell} \cdot \mathbf{C}_1 \cdot \mathbf{u}_{0X}$$

is analogous to the previously encountered nonuniform expansion (8).

Each member of (17) may share the same difficulty. Now, however, the resolution is known. For *each* small parameter we define a *slow* time

$$\tau_1 = \epsilon t,$$

$$\tau_{k+1} = \mu_k t, \qquad k = 1, \ldots, m, \tag{19a}$$

and for clarity write $\tau_0 = t$ for the *fast* time, so that after the transformation (18),

$$\frac{\partial}{\partial t} = \frac{\partial}{\partial \tau_0} + \epsilon \frac{\partial}{\partial \tau_1} + \sum_{k=1}^{m} \mu_k \frac{\partial}{\partial \tau_{k+1}}. \tag{19b}$$

The arbitrary function U is explicitly independent of τ_0 but may now depend upon the τ_j, $j \geq 1$. Equations (17), transformed to coordinates $(X, \tau_0, \tau_1, \tau_{k+1})$, $k = 1, \ldots, m$, are

$$\mathbf{u}_{1_{\tau_0}} + (\mathbf{C}_0 - \lambda \mathbf{I}) \cdot \mathbf{u}_{1_X} = -\mathbf{C}_1 \cdot \mathbf{u}_{0_X} - \mathbf{u}_{0_{\tau_1}}$$

and

$$\mathbf{u}_{(k+1)_{\tau_0}} + (\mathbf{C}_0 - \lambda \mathbf{I}) \cdot \mathbf{u}_{(k+1)_X} = \mathbf{D}_k(\mathbf{u}_c, \mathbf{u}_0, x, t) - \mathbf{u}_{0_{\tau_{k+1}}}.$$

Premultiplication by $\boldsymbol{\ell}$ leads to the same difficulty as before, unless the righthand side is orthogonal to $\boldsymbol{\ell}$; that is, unless

$$\boldsymbol{\ell} \cdot \mathbf{C}_1 \cdot \imath U_X + \boldsymbol{\ell} \cdot \imath U_{\tau_1} = 0,$$

and

$$\boldsymbol{\ell} \cdot \mathbf{D}_k - \boldsymbol{\ell} \cdot \imath U_{\tau_{k+1}} = 0 \qquad k = 1, \ldots, m, \tag{20}$$

where we have substituted for $\mathbf{u}_0$ from (16). Because $U(X, \tau_1, \ldots, \tau_m)$ is arbitrary, we may satisfy the orthogonality conditions (20) by requiring U to be governed by (20). Since the τ_k are related to t, it is frequently more clear to "reconstitute" a single equation governing the time evolution of U. Utilizing (19b), we obtain the reconstituted equation

$$\boldsymbol{\ell} \cdot \imath U_t + \epsilon \boldsymbol{\ell} \cdot \mathbf{C}_1 \cdot \imath U_X = \boldsymbol{\ell} \cdot \left[\sum_{k=1}^{m} \mu_k \, \mathbf{D}_k(\imath U, X, t) \right]$$

whose solution renders $\mathbf{u}_0$ valid through time $O\left(\dfrac{1}{\epsilon}+\dfrac{1}{\mu_1}+\cdots+\dfrac{1}{\mu_m}\right)$ (assuming that all linear quantities in ϵ or μ_i are larger than any quadratic combination of these parameters).

1c. Burgers and Korteweg-deVries Equations

The equations to be derived for dissipative gas dynamics and for shallow-water waves are of the form (12) with operators satisfying the restrictions of Subsection 1b. For example, for the gas dynamics problem,

$$\mathbf{C}_0=\begin{pmatrix}0 & 1 & 0\\ 1 & 0 & 1\\ 0 & 0 & 0\end{pmatrix}, \qquad \mathbf{C}_1(\imath U)=U\begin{pmatrix}1 & 1 & 0\\ \gamma-2 & 1 & \gamma-1\\ 0 & 0 & 1\end{pmatrix}$$

(where γ is the ratio of specific heats, a constant property of the gas). The eigenvalues of $\mathbf{C}_0$ are $+1$, -1, and 0, with corresponding right column eigenvectors with transposes $(1, 1, 0)$, $(1, -1, 0)$, and $(-1, 0, 1)$, respectively. The eigenvalues ± 1 correspond to the speed of sound in the undisturbed medium, and the third eigenvalue corresponds to the convection of "entropy" disturbances with the gas. Sound propagation is the phenomenon of interest. Focusing, therefore, on sound propagating in the direction of increasing x by selecting $\lambda = 1$, we have the corresponding left eigenvector $\boldsymbol{\ell}=(1, 1, 1)$. In the planar case there is only one small parameter μ, in addition to ϵ, and it is given in conventional fluid mechanics notation by $\mu_1 = \mathrm{Re}^{-1}$, an inverse Reynolds number. Corresponding to $\mathbf{D}_1$ we have the operator

$$\mathbf{D}_1(\mathbf{u})=\begin{pmatrix}0 & 0 & 0\\[4pt] 0 & \dfrac{4}{3}+\nu & 0\\[6pt] \dfrac{\gamma-1}{\mathrm{Pr}} & 0 & \dfrac{\gamma}{\mathrm{Pr}}\end{pmatrix}\cdot \mathbf{u}_{xx},$$

where the Prandtl number, Pr, and ratio of shear and dilatational coefficients of viscosity, ν, are assumed constant. Since

$$\boldsymbol{\ell}\cdot\mathbf{D}_1=\left(\frac{4}{3}+\nu+\frac{\gamma-1}{\mathrm{Pr}}\right)U_{XX}, \quad \boldsymbol{\ell}\cdot\imath=2, \text{ and } (\boldsymbol{\ell}\cdot\mathbf{C}_1\cdot\imath)=(\gamma+1)U,$$

equations (20) are

$$U_{\tau_1} + \frac{1}{2}(\gamma + 1)UU_X = 0$$

and

$$U_{\tau_2} = \frac{1}{2}\left(\frac{4}{3} + \nu + \frac{\gamma - 1}{\text{Pr}}\right)U_{XX},$$

and the reconstituted equation is Burgers equation

$$U_t + \epsilon \frac{\gamma + 1}{2} UU_X = \frac{1}{2}\left(\frac{4}{3} + \nu + \frac{\gamma - 1}{\text{Pr}}\right)\text{Re}^{-1}U_{XX}.$$

Shallow-water waves will be shown to be governed by a set of two equations (equivalent to the approximation of Boussinesq) for which

$$\mathbf{C}_0 = \begin{pmatrix} 0 & 1 \\ 1 & 0 \end{pmatrix}, \qquad \mathbf{C}_1(\mathbf{u}) = \begin{pmatrix} u_2 & u_1 \\ 0 & u_2 \end{pmatrix}.$$

The equations depend upon ϵ and a single small parameter μ_1 representing the circumstances of the propagation and the operator $\mathbf{D}_1$ is

$$\mathbf{D}_1(\mathbf{u}) = \begin{pmatrix} \frac{1}{3}u_{2xxx} \\ u_{2xxt} \end{pmatrix}.$$

The eigenvalues of $\mathbf{C}_0$ are $\lambda = 1$, $\lambda = -1$, corresponding to the right column vectors with transposes $(1, 1)$ and $(1, -1)$ and to the left (row) eigenvectors $(1, 1)$ and $(1, -1)$. Thus, selecting $\lambda = 1$, with $\boldsymbol{\ell} \cdot \boldsymbol{\imath} = 2$, $\boldsymbol{\ell} \cdot \mathbf{D}_1 = -\frac{2}{3}U_{XXX}$, $\boldsymbol{\ell} \cdot \mathbf{C}_1 \cdot \boldsymbol{\imath} = 3U$, and the appropriate form of equations (20) are

$$U_{\tau_1} + \tfrac{3}{2}UU_X = 0$$

and

$$U_{\tau_2} + \tfrac{1}{3}U_{XXX} = 0,$$

and the reconstituted equation is the Korteweg-deVries equation

$$U_t + \tfrac{3}{2}\epsilon UU_X + \tfrac{1}{3}\mu_1 U_{XXX} = 0.$$

2. Dissipative Gas Dynamics and the Burgers Equation

The equations of gas dynamics constitute a quasilinear hyperbolic system. Such systems are discussed in the previous chapter. The unsteady equations of inviscid fluid mechanics are "properly embedded" in the full viscous equations. One of the remarkable equations of fluid dynamics, the Burgers equation, illustrates this embedding. It describes the motion of weak nonlinear waves in gases when a first accounting of dissipative effects is needed. In the limit of vanishing dissipation, this equation provides the proper interpretation for the inviscid solution. The history of this equation is too rich to detail here. It was proposed by Burgers (1948) as a model equation for one-dimensional turbulence. Lighthill (1956) showed that, with the proper interpretation, the equation was the appropriate one for weak planar wave propagation.

Here Burgers equation is derived for one-dimensional unsteady gas motions with planar, cylindrical, or spherical symmetry. The space coordinate will be denoted by r. Our interest will be in motions of small amplitude characterized by a parameter ϵ, and the density ρ; the velocity $\hat{v}$ (speed in the r-direction) and the entropy $\hat{s}$ are made dimensionless as follows:

$$\hat{\rho} = \rho_*(1 + \epsilon\rho), \quad \hat{v} = \epsilon a_* v, \quad \hat{s} = s_* + \epsilon c_p \sigma.$$

The subscript $(\;)_*$ indicates ambient conditions that we take to be uniform; a is the sound speed and c_p the specific heat at constant pressure. The appropriate nondimensionalization of the independent variables is one that measures distances in units of a typical wavelength l and time in units of the time it takes the wave to propagate the distance l:

$$r = xl, \qquad \hat{t} = \frac{l}{a_*} t.$$

An appropriate measure of l is $l = (\hat{v})_{max}/(\hat{v}_r)_{max}$, which characterizes the maximum slope of the velocity profile. (A sinusoidal wave has a wavelength of $2\pi l$.) With this notation the unsteady Navier-Stokes equations are

$$\mathbf{u}_t + \mathbf{C}(\mathbf{u}) \cdot \mathbf{u}_x = \mathrm{Re}^{-1}[\mathbf{D}_1(\mathbf{u}) + O(x_p^{-2}\mathbf{u}, x_p^{-1}\mathbf{u}_x, \epsilon\mathbf{u}_x^2)] + x_p^{-1}\mathbf{D}_2(\mathbf{u}, x), \quad (21)$$

where $\mathbf{u}$ is the column vector of unknowns and $\mathbf{D}_2$ is a column vector that depends upon the geometry,

$$\mathbf{u} = \begin{pmatrix} \rho \\ v \\ \sigma \end{pmatrix}, \qquad \mathbf{D}_2(\mathbf{u}) = -j\frac{x_p}{x}\begin{pmatrix} [1+\epsilon\rho]v \\ 0 \\ 0 \end{pmatrix},$$

with $j = 0$, 1, 2 for planar, cylindrical, and spherical motions and where $\mathbf{C}$ and $\mathbf{D}_1$ are

$$
\mathbf{C}(\mathbf{u}) = \begin{pmatrix} \epsilon v & 1 + \epsilon \rho & 0 \\ \dfrac{1 + \epsilon T}{1 + \epsilon \rho} & \epsilon v & 1 + \epsilon T \\ 0 & 0 & \epsilon v \end{pmatrix}, \qquad \mathbf{D}_1(\mathbf{u}) = \begin{pmatrix} 0 & 0 & 0 \\ 0 & \dfrac{4}{3} + v & 0 \\ \dfrac{\gamma - 1}{\mathrm{Pr}} & 0 & \dfrac{\gamma}{\mathrm{Pr}} \end{pmatrix} \cdot \mathbf{u}_{xx}.
$$

In $\mathbf{C}$, T is dimensionless temperature and is related to σ and ρ by

$$
\exp(\epsilon \gamma \sigma) = \frac{1 + \epsilon T}{(1 + \epsilon \rho)^{\gamma - 1}}.
$$

In addition to the small parameter ϵ characterizing the amplitude of the wave, we have the small parameters $x_p^{-1} = l/r_p$ characterizing the initial motion, which may be thought of as being generated by a piston motion $r_p(t)$, where $r_p(t) \ll a_*$, and $\mathrm{Re}^{-1} = \mu_*/\rho_* a_* l$, where μ_* is the viscosity and Re^{-1} is the ratio of viscous to inertial forces. Air at standard sea-level conditions has a Reynolds number, Re, of about $2 \cdot 10^7 l$ where l is measured in meters. The other parameters occurring in (21), the ratio of the specific heats γ, the ratio of the shear and dilatational coefficients of viscosity v, and the Prandtl number Pr, are all $O(1)$. For air, $\gamma = 1.4$ and $\mathrm{Pr} = 0.72$ under fairly diverse conditions, while v, which accounts for a loss of translation energy to molecular vibrations in polyatomic gases, varies markedly with the water-vapor content. We have not written out the full righthand side of (21) as the terms represented only by their order are eliminated by our approximation.

2a. The Burgers Equation

Our concern is the behavior and structure of the wave motions governed by (21). In particular, we limit our interest to the simplest possible wave motions that embody the principal features of (21), the nonlinearity in the $\mathbf{C}(\mathbf{u}) \cdot \mathbf{u}_x$ term and the effect of the very small righthand side. Consequently, we consider the following model problems. A planar, cylindrical or spherical piston located at $r = r_p \gg l$ undergoes a small-amplitude motion $\epsilon r_p(t)$. The small parameter x_p^{-1} occurring in

(21) is l/r_p. This motion gives rise to a disturbance that travels, to the first approximation, outward at the ambient speed of sound. We ask how this wave evolves with time for all subsequent times.

The system (21) is of the form (12) with $\mu_1 = \mathrm{Re}^{-1}$ and $\mu_2 = x_p^{-1}$. Applying the procedure of Section 2, we find $\mathbf{u}_0 = U(x - \lambda t)\mathbf{\iota}$. There are three distinct eigenvalues for our $\mathbf{C}_0$: ± 1 and 0; they have the column eigenvectors with transposes $(1, \pm 1, 0)$ and 0 representing, to lowest order, the two sound-speed modes and the single entropy mode. For the model problem a proper mode is sound propagation in the positive direction. Thus with $\lambda = +1$ we have

$$\begin{pmatrix} \rho \\ v \\ \sigma \end{pmatrix} = U(x - t) \begin{pmatrix} 1 \\ 1 \\ 0 \end{pmatrix}, \tag{22}$$

where U is determined by the piston motion. The resultant motion is planar propagation at the ambient sound speed unaffected by non-linearity or diffusion. Naturally, we cannot expect this result to apply far from the piston; we know that the amplitude will depend critically on the geometry there, and, as indicated earlier, higher-order nonlinear effects will accumulate to make a first-order contribution. This contribution causes the wave to steepen and brings viscous dissipation into play.

Before we proceed with the formal approach of Section 1, we note that the final result, the Burgers equation, is almost an immediate consequence of (21) if we take the new coordinates to be $\xi = x - t$ and $\tau = \epsilon t$. Then, with $\lambda = +1$, (21) becomes

$$[\mathbf{C}(\mathbf{u}) - \mathbf{I}] \cdot \mathbf{u}_\xi + \epsilon \mathbf{u}_\tau = \mathrm{Re}^{-1} \left[\mathbf{D}_1(\mathbf{u}) + O\left(\epsilon^2 \frac{u}{\tau^2}, \epsilon \frac{u_\xi}{\tau}, \epsilon u_\xi^2 \right) \right] + x_p^{-1} \mathbf{D}_2(\mathbf{u}, \xi). \tag{23}$$

Because the operator $\mathbf{C}_0 - \mathbf{I}$ is singular, the lowest-order system contained in (23) is redundant:

$$-\rho_\xi + v_\xi = O(\epsilon, \mathrm{Re}^{-1}),$$

$$\rho_\xi - v_\xi + \sigma_\xi = O(\epsilon, \mathrm{Re}^{-1}),$$

and

$$-\sigma_\xi = O(\mathrm{Re}^{-1}).$$

The first two equations tell us that $\rho = v$, and the redundancy is eliminated by summing all three equations. The resulting sum of the righthand sides with $\rho = v$ comprises the other equation for ρ and v:

$$v_\tau + \left(\frac{\gamma+1}{2}\right)vv_\xi + \frac{jv}{2\tau} = \frac{\text{Re}^{-1}}{2\epsilon}\left(\frac{4}{3} + v + \frac{\gamma-1}{\text{Pr}}\right)v_{\xi\xi}. \tag{24}$$

This is the Burgers equation generalized to apply to motions with cylindrical or spherical symmetry.

Returning to the general scheme, we introduce the slow times according to (19a). With $x = X + t = X + x_p\tau_3$ and noting that

$$\boldsymbol{\ell}\cdot\mathbf{C}_1\cdot\boldsymbol{\imath} = (\gamma+1)U(X), \quad \boldsymbol{\ell}\cdot\mathbf{D}_1 = \left(\frac{4}{3} + v + \frac{\gamma-1}{\text{Pr}}\right)U(X)_{XX},$$

$$\boldsymbol{\ell}\cdot\mathbf{D}_2 = -j[\tau_3 + x_p^{-1}X]^{-1} \quad \text{and} \quad \boldsymbol{\ell}\cdot\boldsymbol{\imath} = 2,$$

we find the three equations

$$U_{\tau_1} + \Gamma U U_X = 0, \tag{25a}$$

$$U_{\tau_2} - \frac{\kappa}{2}U_{XX} = 0, \tag{25b}$$

and

$$U_{\tau_3} + \frac{j}{2}\frac{U}{\tau_3} = 0, \tag{25c}$$

where we have set $\Gamma = (\gamma+1)/2$, $\kappa = \frac{4}{3} + v + (\gamma-1)/\text{Pr}$ and dropped $x_p^{-1}X$ relative to τ_3.

This is a "set" of three equations for U that apply sequentially to U depending upon the comparative sizes of the time scales defined by the three τ's. The physical time scales are

$$\hat{t}_1 = \frac{l}{(\dot{r}_p)_{max}}, \quad \hat{t}_2 = \frac{\rho_*l^2}{\mu_*} \quad \text{and} \quad \hat{t}_3 = \frac{r_p}{a_*},$$

and they correspond to: (1) the time it takes a disturbance to propagat a distance l at the maximum perturbation velocity; (2) the time it take a disturbance to diffuse a distance l; and (3) the time it takes the wave t propagate a distance r_p. The reconstituted equation corresponding to the set (25) is

$$U_t + \epsilon \Gamma U U_X + \frac{j}{2t} U = \frac{\kappa}{2} \text{Re}^{-1} U_{XX}, \qquad (26)$$

which may be compared to (24) with $\tau = \epsilon t$.

2b. Evolution of the Waveform

The manner in which a given waveform evolves depends upon the ordering of the times $\hat{t}_1$, $\hat{t}_2$, $\hat{t}_3$. If the time $\hat{t}_2$ is substantially less than $\hat{t}_1$, then the wave will decay sufficiently rapidly and nonlinear effects are never important. Superimposed on this viscous decay will be the additional reduction in amplitude due to geometrical dilatation. Regardless of which of the two times, $\hat{t}_3$ and $\hat{t}_2$, is the smaller, the asymptotic behavior will involve both. Thus, from the second and third of equations (25), taken in either order, we find

$$U = (\tau_3)^{-j/2} f(X, \tau_2), \qquad (27)$$

where f satisfies (25b). In this particular case, the solution (27) to (25b) and (25c) is identical to solutions of the reconstituted linear equation

$$\frac{\partial U}{\partial t} + j \frac{U}{2t} = \text{Re}^{-1} \frac{\kappa}{2} \frac{\partial^2 U}{\partial X^2}. \qquad (28)$$

We note that the geometric and diffusive effects are essentially separable in this equation, as our multi-time procedure illustrates naturally in (27).

Our primary concern is with nonlinear wave phenomena. As we have noted, if $\hat{t}_2 \ll \hat{t}_1$, then the solution is given by (27), and nonlinear effects are uniformly of higher order. However, if $\hat{t}_1 \leq \hat{t}_2$, then nonlinear effects are important for nearly all times, and the correct time scale for diffusive effects is, as we shall see, no longer $\hat{t}_2$ but related to $\hat{t}_1$.

As we might expect from (25), the geometrical and nonlinear effects are also separable, and solving the first two equations in either order gives the same result. For example, with $\hat{t}_3 \ll \hat{t}_1$ we find from (25c)

$$U(X, \tau_1, \tau_3) = \tau_3^{-j/2} F(X, \tau_1).$$

The first equation then requires

$$F_{\tau_1} + \tau_3^{-j/2} \Gamma FF_X = 0.$$

This equation has the (implicit) general solution

$$F = \mathscr{F}(X - \Gamma F \int \tau_3^{-j/2} \, d\tau_1),$$

where $\mathscr{F}$ is an arbitrary function of its argument. This follows directly from the method of characteristics for first-order equations and can be verified by direct substitution. On this longer time scale we must account for the variability of τ_3. Conversely, for $\hat{t}_1 \ll \hat{t}_3$, (25a) gives

$$U(X, \tau_1, \tau_3) = g(X - \Gamma \int U \, d\tau_1, \tau_3),$$

where g is an arbitrary function of its argument and we have anticipated that U will depend on τ_3. Because we are interested in changes over times comparable to τ_3, the quadrature with respect to τ_1 does not give $U\tau_1$ but must account for the variation of τ_3 with τ_1. Now (25c) implies that $U\tau_3^{j/2}$ is independent of τ_3; thus

$$U = g = \tau_3^{-j/2} h(X - \Gamma U\tau_3^{j/2} \int \tau_3^{-j/2} \, d\tau_1).$$

Consequently, in either case we find that for times comparable to the longer of $\hat{t}_1$, $\hat{t}_3$ that

$$U(X, t) = \left(\frac{t_0}{t}\right)^{j/2} \mathscr{F}\left(X - \epsilon \Gamma U t^{j/2} \int_{t_0}^{t} \frac{dt}{t^{j/2}}\right), \tag{29}$$

where $\mathscr{F}$ is determined by the conditions at $t = t_0$: $U(X, t_0) = \mathscr{F}(X)$. This is, of course, the solution to the combined equation

$$\frac{\partial U}{\partial t} + j \frac{U}{2t} + \epsilon \Gamma U \frac{\partial U}{\partial X} = 0 \tag{30}$$

that governs motion for $\hat{t}_3$, $\hat{t}_1 \ll \hat{t}_2$ when $\hat{t}_3$ and $\hat{t}_1$ are comparable.

We have scaled the coordinates X and t so that X is measured in terms of the maximum gradients in U. Thus, according to (29), U would become multivalued in a time such that $\epsilon \Gamma U t^{j/2} \int t^{-j/2} \, dt$ is comparable to unity with $U t^{j/2}$ fixed, or when

$$\epsilon t^{1-(j/2)} [\log t]^{j(j-1)/2} = O(1),$$

if this steepening were not resisted by the viscous diffusion represented by the second of equations (25). Thus, that equation becomes effective on a time scale $\hat{t}'_2$ given by

$$\hat{t}'_2 = \begin{cases} \hat{t}_1, & j=0, \\ \hat{t}_1/\epsilon, & j=1, \\ \hat{t}_1\epsilon \exp\dfrac{1}{\epsilon}, & j=2. \end{cases}$$

This time reflects more accurately the time required for nonlinear, and consequently diffusive, effects to become important in that it accounts for the reduction in amplitude due to geometric dilatation. With $\epsilon = 0.1$ the times for a wave to steepen are $\hat{t}_1$, $10\hat{t}_1$, and $2\cdot 10^3\hat{t}_1$, respectively; for $\epsilon = 0.01$ they are $\hat{t}_1$, $100\hat{t}_1$, and $2.7\cdot 10^{41}\hat{t}_1$. Thus diffusive effects come to the fore on the shorter of the two time scales $\hat{t}_2'$, $\hat{t}_2$. If $\hat{t}_2 \gg \hat{t}'_2$, however, (26) indicates that we need only account for viscous diffusion in small regions about the steepened wavefront.

On time scales comparable to the largest of $\hat{t}_1$, $\hat{t}_2'$, $\hat{t}_3$, all of equations (25) must be taken into account unless $\hat{t}_2 \ll \hat{t}_2$, the case we considered first. Thus with $\hat{t}_2 \le \hat{t}_2$ we see that the motions must be governed by (26). With $t = \tau/\epsilon\Gamma$ and $\delta = \kappa/(\epsilon\,\mathrm{Re})$, (26) becomes

$$U_\tau + \frac{j}{2\tau}\,U + UU_X = \frac{\delta}{2}\,U_{XX}. \tag{31}$$

In the limit $\delta \to 0$ the solution to (31) is given by the equivalent of (29),

$$\tau^{j/2}U(X,\,\tau) = F\left(X - U\tau^{j/2}\int_{\tau_0} \tau^{-j/2}\,d\tau\right). \tag{32}$$

The multivaluedness of (32) is resolved by introducing "discontinuities" —that is, shock waves. For any $\delta > 0$ these discontinuities must have a continuous structure determined by a balance between X derivatives in (31). The dimensionless length scale of this structure is obviously $O(\delta)$. Thus the prescription (32), with the attendant discontinuities, provides the correct physical description of the solution to (32) as $\delta \to 0^+$, and (30), with $\tau = \epsilon\Gamma t$, is properly embedded in the more complete model (31). (See Subsection 4d of Chapter III.) The changes or jumps in the physical variables across the discontinuities can be determined by

solving (31) and then extracting the limit of the resulting solution as
$\delta \to 0$.

An alternative approach, to be explored here, is to deal with the
hyperbolic limit equation obtained from (31) by putting $\delta = 0$. In this
approach, one must recognize that *the associated conservation laws must
be consistent with those of the full equation if the limit equation is to be
properly embedded in a more complete model.* For disturbances that
vanish sufficiently rapidly as $|X| \to \infty$, the conservation law may be
found by integrating (31) over all X. The result is

$$\tau^{j/2} \int_{-\infty}^{\infty} U(X, \tau) \, dX = const. \tag{33}$$

Thus if our initial waveform is $F(\xi)$, then at subsequent times it will
distort according to (32), as shown in Figure IV.1. The multivaluedness

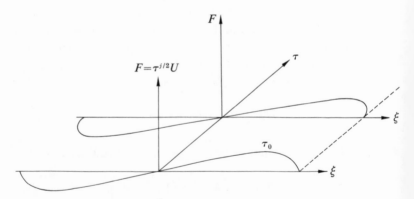

IV.1. Distortion of initial waveform by nonlinear effects.

is resolved by introducing discontinuities that preserve the area under
the curve according to (33).

2c. Structure of Weak Shocks

The structure of the discontinuities, or shock waves, can be determined by moving with them at their unknown speeds, so long as their
thickness is much smaller than the length scale that characterizes
gradients in the flow ahead or behind them. Consequently, we introduce

a coordinate $\zeta = (X - \int \mathscr{S} \, d\tau)/\delta$ that expands the X scale to one appropriate to the wave's structure and that moves with the unknown shock speed $\mathscr{S}$. This gives

$$- \mathscr{S} U_\zeta + U U_\zeta - \tfrac{1}{2} U_{\zeta\zeta} = \delta \left[U_\tau + \frac{j}{2\tau} U \right],$$

and we note that to lowest order the shock structure is independent of τ. Assuming that U vanishes in front of the shock ($\zeta \to \infty$) and that it tends to the (time-varying) value U_s behind the shock ($\zeta \to -\infty$), a first quadrature shows that $\mathscr{S} = U_s/2$. A second quadrature gives

$$U = U_s[1 + \exp(U_s\zeta)]^{-1} = \tfrac{1}{2}U_s[1 - \tanh \tfrac{1}{2}(U_s\zeta)]. \tag{34}$$

This structure, which in terms of dimensional variables moves with the speed $a_* + (\Gamma/2)\hat{v}_s$, where $\hat{v}_s$ is the flow speed behind the shock, has a characteristic "maximum slope" thickness $\Delta = \hat{v}_s/(\hat{v}_r)_{max}$ given by

$$\mathrm{Re}_{\hat{v}_s,\Delta} = \frac{\rho_* \hat{v}_s \Delta}{\mu_*} = 4\kappa, \tag{35}$$

which, for air at sea level, varies from about 5 to 50 times $(a_*/\hat{v}_s)10^{-7}$ meters. The result (34) was first given by Taylor (1910).

2d. Asymptotic Behavior of the Waveform

So long as the thickness of the shock is much smaller than the distance over which the flow outside the shock changes significantly, we may treat the shock as a discontinuity. It follows naturally, then, that waves will distort as shown in Figure IV.1, producing a sawtooth wave or, in the case depicted, an "N" shaped wave when shock waves are inserted to make the function single valued. The subsequent behavior of this sawtooth or N-wave is simply a geometric proposition until the wave becomes so weak that the structure of the shock is important. Thus, for much of its evolution the wave distorts in the manner illustrated in Figure IV.2. The wave will have triangular elements like that shown for τ_0. At a later time τ the element would be a longer but triple-valued triangular pulse with the same amplitude. A discontinuity is introduced to make the function $F = \tau^{j/2}U$ into the single-valued triangular pulse of length L with reduced amplitude. We note that the balance (33) requires

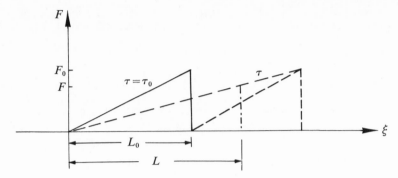

IV.2. Asymptotic behavior of N-wave (Re $\gg$ 1).

that the triangular areas $L_0 F_0/2$ and $LF/2$ be equal; geometry requires

that F_0: $L_0 + F_0 \int_{\tau_0}^{\tau} t^{-j/2}\, dt = F$:$L$, and, consequently,

$$\frac{F}{F_0} = \left(\frac{\tau}{\tau_0}\right)^{j/2} \frac{U}{U_0} \frac{L_0}{L} = \left[1 + \frac{F_0}{L_0} \int_{\tau_0}^{\tau} t^{-j/2}\, dt\right]^{-1/2}.$$

For large τ we have

$$\frac{U}{U_0} \propto \tau^{-1/2-j/4} (\log \tau)^{-j(j-1)/4}, \quad \text{and} \quad \frac{L}{L_0} \propto \tau^{1/2-j/4} (\log \tau)^{j(j-1)/4} \quad (36)$$

so long as the shock-wave thickness is small compared to L. Thus the shock-wave strength decreases as $\tau^{-1/2}$, $\tau^{-3/4}$, and $\tau^{-1} (\log \tau)^{-1/2}$, while the overall wavelength increases as $\tau^{1/2}$, $\tau^{1/4}$, and $(\log \tau)^{1/2}$ for $j=0, 1$, and 2.

Except in the case of one-dimensional motion with only a positive phase ($U \geq 0$), the shock structure will eventually become so large that the shock thickness is comparable to the length L based on $(\hat{v}_r)_{max}$ behind the shock. During this stage all the terms in (31) are in balance over most of the wave, and the results (36) no longer apply. After this stage, the nonlinear terms are no longer of consequence, and the final asymptotic decay follows from (28):

$$U \sim \left(\frac{\tau_0}{\tau}\right)^{(j+1)/2} \frac{A_0}{2\delta} \frac{X}{\tau} \exp\left(\frac{-X^2}{2\delta\tau}\right), \quad (37)$$

where A_0 is the area $\displaystyle\int_{-\infty}^{\infty} U(X, \tau_0)\, dX$. Here we have assumed that the initial conditions lead to a balanced N-wave, that is, that they correspond to a dipole.

The transition to this behavior from that given by (36) occurs when the length of the wave is comparable to the shock thickness—that is, when the Reynolds number based on the velocity behind the shock and the wavelength is comparable to that given by (35), $\mathrm{Re}_{v_{s,\,l}} \approx 4\kappa$. A convenient definition of the Reynolds number for an arbitrary pulse is the "lobe" Reynolds number R defined by

$$R = \frac{\rho_*}{\mu_*} \int_{r_1}^{r_2} \hat{v}(r, \hat{t})\, dr = \epsilon \, \mathrm{Re} \int_{X_1}^{X_2} U(X, \tau)\, dX,$$

where r_1, r_2 are adjacent zeros of $\hat{v}$ and X_1, X_2 are adjacent zeros of $U(X, \tau)$. Integrating (31) between successive zeros we find that

$$\frac{d}{d\tau}\left(R\tau^{j/2}\right) = \tau^{j/2}\,\frac{\kappa}{2}\,[U_X(X_2, \tau) - U_X(X_1, \tau)]. \tag{38}$$

A pulse with only a single phase—that is, with $U \geq 0$ (or $U \leq 0$) everywhere, has $U_X(X_{1,\,2}, \tau) = 0$ and $R \propto \tau^{-j/2}$. In this case, with $j = 0$, we see that R is constant and there is no transition to the viscous acoustic behavior described by (37). When the pulse is headed by a shock moving into a quiescent medium, we have $U_X(X_2, \tau) = 0$, and may conclude from Figure IV.2 that

$$U_X(X_1, \tau) = U_X(X_1, \tau_N)\,(\tau_N/\tau)^{j/2}\left[1 + U_X(X_1, \tau_N)\tau_N^{j/2}\int_{\tau_N}^{\tau} t^{-j/2}\,dt\right]^{-1}$$

for times $\tau \geq \tau_N$, where τ_N is some time at which the wave is of the half N-wave form shown. Combining this result with (38), we may determine the value of R at subsequent times; for $\tau \gg \tau_N$ we find, for $R \gg 1$,

$$R - R_N\left(\frac{\tau_N}{\tau}\right)^{j/2} \approx -\frac{\kappa}{2}\begin{cases} \log\,(\tau/\tau_N) & j = 0, \\ 1 - (\tau_N/\tau)^{1/2} & j = 1, \\ \tau^{-1}\displaystyle\int_{\tau_N}^{\tau} d\tau/\log \tau & j = 2, \end{cases} \tag{39}$$

where R_N is the lobe Reynolds number when $\tau = \tau_N$. The lobe Reynolds decreases from R_N according to (39) until R is no longer $\gg 1$. Eventually there is a transition to viscous acoustic behavior, and, using (37) to determine $U_X(X_1, \tau)$ in (38), we find $\tau^{j/2}R = const. + \kappa\tau^{-1/2}$. The linear solution shows that the constant must be zero, so that ultimately

$$R \sim R_0\left(\frac{\tau_0}{\tau}\right)^{(1+j)/2}. \tag{40}$$

The result (39), based upon the assumption that $R \gg 1$, has been shown by numerical computation for $j = 1$ and 2 to hold for R's down to one and below [Sachdev and Seebass (1973)].

2e. Hopf-Cole Transformation

Our study of weak one-dimensional waves has provided a fairly complete picture of their behavior for all times except those when the wave is in transition from the moderate Reynolds number description of (36) to the small Reynolds number diffusion of (37). Fortunately, for the one-dimensional case we can supply these details by utilizing the Hopf-Cole transformation, $\psi = \exp\left(\frac{1}{\delta}\int_X^{\infty} U dX\right)$, which reduces (31) to the heat equation for ψ when $j = 0$†. Thus we can solve the initial-value problem to find

$$U(X, \tau) = \frac{\displaystyle\int_{-\infty}^{\infty} \frac{X - \xi}{\tau} \exp\left\{\frac{1}{\delta}\left[\int_{\xi}^{\infty} U(y, 0)\, dy - \frac{(X - \xi)^2}{2\tau}\right]\right\}d\xi}{\displaystyle\int_{-\infty}^{\infty} \exp\left\{\frac{1}{\delta}\left[\int_{\xi}^{\infty} U(y, 0)\, dy - \frac{(X - \xi)^2}{2\tau}\right]\right\}d\xi},$$

which provides the detailed structure for the wave for all time. The behavior of the lobe Reynolds number corresponding to (39) and (40) is found to be (see Lighthill, 1956)

$$R = \kappa \log\left\{1 + \left(\frac{\tau_N}{\tau}\right)^{1/2}\left[\exp\left(R_N/\kappa\right) - 1\right]\right\}$$

in agreement with our predictions (39) and (40), but supplying details not available to us without a solution to (31). This solution vindicates

† For $j \neq 0$ the result is

$$\psi_t + \frac{j}{2t}\psi \log \psi = \frac{\delta}{2}\psi_{XX}.$$

all of our results (for $j=0$) and provides a remarkably exact description of a complex nonlinear physical phenomenon. Some of the mathematical complexity that arises in the other geometries can be readily seen by writing $V = U\tau^{j/2}$ and $T = \int \tau^{-j/2} \, d\tau$. With this transformation (31) becomes

$$V_T + VV_X = \frac{\delta}{2} g(T) V_{XX},$$

where $g(T) = \frac{1}{2}T$ for $j=1$ and $g(T) = \exp T$ for $j=2$. This equation can serve as an alternate vehicle for the derivation of our results for the cases $j = 1$ or 2, but obscures the basic physics indicated so clearly by (25c).

3. Shallow-Water Waves and the Korteweg-deVries Equation

Water wave problems usually emanate from a class of equations (elliptic) generally thought of as antithetic to propagation phenomena. Water wave propagation is a surface phenomenon, arising not from the nature of the field equations, but rather in response to boundary conditions. Laplace's equation describes fluid motion in the absence of viscosity and rotational body forces, providing that the initial disturbance is irrotational. With these assumptions, and assuming further that the motion is two-dimensional, the equations that we need to consider are Laplace's equation

$$\Phi_{xx} + \Phi_{yy} = 0 \tag{41}$$

and the Bernoulli equation

$$\Phi_t + \tfrac{1}{2}(\Phi_x^2 + \Phi_y^2) + \frac{p}{\rho} + gy = F(t) \tag{42}$$

that relates the velocity potential Φ, the pressure p (divided by the density ρ), and the gravitation acceleration g, which acts in the negative y direction. The undisturbed free surface is located at a height $y = h$ above the solid bottom $y = 0$, $-\infty < x < \infty$. The function $F(t)$ in (42) is arbitrary and may be absorbed in the definition of Φ. The velocity vector $(u, v) = \text{grad } \Phi$.

In the absence of viscous and surface tension effects, the appropriate boundary conditions are

$$\Phi_y(x, 0, t) = 0 \tag{43}$$

(no flow through the bottom), while at the free surface $y = y^*(x, t)$ the pressure is constant, and the kinematic requirement that fluid particles at the free surface always remain there must be imposed. The latter two conditions may be expressed by the equations

$$p(x, y^*(x, t), t) = 0$$

and

$$y^*_t + u^* y^*_x = v^*,$$

where the asterisk identifies functions evaluated at $y = y^*(x, t)$. When the pressure is first evaluated at the free surface and then differentiated by x, a multiple of the derivative along the free surface is obtained and must vanish. Substitution for p/ρ from the Bernoulli equation produces the condition

$$\Phi^*_t + \frac{1}{2}(u^{*2} + v^{*2}) + gy^* = 0. \tag{44}$$

3a. Infinitesimal Waves

Suppose the free surface is disturbed by an infinitesimal amount, so that

$$y^*(x, t) = h[1 + \epsilon\eta(x, t)]$$

and $\epsilon \to 0$. Then it is clear that Φ is $O(\epsilon)$, and so it is convenient to put $\Phi = \epsilon\phi$. Function values for any quantity (say f) at the free surface are replaced by its Taylor expansion about the undisturbed free surface $\eta = 0$, which is denoted by a zero subscript; that is,

$$f(x, y^*, t) = f(x, h, t) + f_y(x, h, t)\epsilon h\eta + \cdots = f_0 + (f_y)_0 \epsilon h\eta + \cdots.$$

Upon division by ϵ, the field equation is

$$\phi_{xx} + \phi_{yy} = 0,$$

and the kinematic and pressure conditions are, respectively,

$$h\eta_t - (\phi_y)_0 = \epsilon h[(\phi_{yy})_0\eta - (\phi_x)_0\eta_x] + O(\epsilon^2) \tag{45}$$

and

$$(\phi_t)_0 + gh\eta = -\epsilon\left[\frac{(\phi_x)_0{}^2 + (\phi_y)_0{}^2}{2} + h\eta(\phi_{yt})_0\right] + O(\epsilon^2). \tag{46}$$

Infinitesimal waves are described by putting $\epsilon = 0$ in (42), (45), and (46), giving

$$\phi_{xx} + \phi_{yy} \qquad = 0; \qquad (47a)$$

$$h\eta_{lt} - \phi_y(x, h, t) = 0; \qquad (47b)$$

$$\phi_t(x, h, t) + gh\eta = 0; \qquad (47c)$$

$$\phi_y(x, 0, t) \qquad = 0. \qquad (47d)$$

By separating variables in the form $\eta = \eta_0 e^{i(kx-\omega t)}$, $\phi = \chi(y)e^{i(kx-\omega t)}$, and substituting into these equations, we easily find the solution

$$\chi = D \cosh ky,$$

$$\eta_0 = \frac{i\omega}{gh} D \cosh kh,$$

where D is an arbitrary (complex) constant, provided that the equation

$$\begin{pmatrix} i\omega h & k \sinh kh \\ igh & \omega \cosh kh \end{pmatrix} \begin{pmatrix} \eta_0 \\ D \end{pmatrix} = 0$$

is satisfied. This is possible only if the determinant of the matrix of coefficients vanishes, thus leading to the dispersion relation

$$\omega^2 = gk \tanh kh. \qquad (48)$$

Water waves are therefore dispersive.

3b. The Long Wave Approximation

Water is considered to be "shallow" for a particular wave problem if the length of the wave l under consideration is great compared to the depth h, or $h/l \ll 1$. Since $l = 2\pi/k$, water is shallow if $kh \ll 1$. Under these circumstances the dispersion relation (48) is well approximated by the first few terms of its Taylor series expansion,

$$\omega^2 = ghk^2[1 - \tfrac{1}{3}(hk)^2 + O[(hk)^4]].$$

As $kh \to 0$, the phase speeds $c = \omega/k$ are given by

$$c = \pm \sqrt{gh},$$

which therefore represent the propagation speeds of waves of great length. In this approximation the dispersion relation is $\omega^2 = ghk^2$, and the limiting case is not dispersive. If, however, account is taken of the effects of small but nonzero hk, the phase speeds are

$$c = \frac{\omega}{k} = \pm \sqrt{gh} \{1 - \tfrac{1}{6}(hk)^2 + O[(hk)^4]\}.$$

Neglecting the term in ω of order $(hk)^5$, the dispersion relation is seen to correspond (see Chapter V, p. 140) to the differential equation

$$\psi_t \pm \sqrt{gh} \left\{ \psi_x + \frac{h^2}{6} \psi_{xxx} \right\} = 0,$$

which is a linearized version of the Korteweg-deVries equation.

The idea now is to take account of nonzero amplitude ($\epsilon \neq 0$) as well as nonzero wavenumber ($kh \neq 0$), by "expanding", in a sense, about the solution valid for $\epsilon = 0$ and $kh = 0$. This was first done by Korteweg and deVries (1895). To do this, it is appropriate to make all quantities dimensionless by using the "natural" physical scales corresponding to the linearized case. Take

$$y = hY, \quad x = l\xi, \quad t = l(gh)^{-1/2}\tau, \quad \mu = \frac{h}{l},$$

and expand Φ in the form

$$\Phi = \epsilon l \sqrt{gh} \{\bar{\phi}_0(Y)F_0(\xi, \tau) + \mu^2 \bar{\phi}_2(Y)F_2(\xi, \tau) + \mu^4 \bar{\phi}_4(Y)F_4(\xi, \tau) + O(\mu^6)\}.$$

Laplace's equation in the scaled coordinates is

$$\Phi_{YY} + \mu^2 \Phi_{\xi\xi} = 0.$$

Substitution of the series shows that the solutions for the $\phi_{2n}(Y)$ that satisfy the boundary condition $\Phi_y(x, 0, t) = 0$ are

$$\bar{\phi}_{2n} = (-1)^n \frac{Y^{2n}}{(2n)!},$$

and the F_{2n} satisfy

$$F_{2n} = \frac{\partial^{2n} F_0}{\partial \xi^{2n}}.$$

Defining $V(\xi, \tau) \equiv F_{0\xi}$ and applying the boundary conditions (45) and (46), one obtains the equations

$$\eta_\tau + V_\xi + \epsilon(V\eta)_\xi = \tfrac{1}{6}\mu^2 V_{\xi\xi\xi} + O(\epsilon\mu^2) \tag{49a}$$

and

$$V_\tau + \eta_\xi + \epsilon V V_\xi = \tfrac{1}{2}\mu^2 V_{\xi\xi\tau} + O(\epsilon\mu^2), \tag{49b}$$

which are equivalent to the approximation of Boussinesq (1872) for shallow water.

We now apply the method of Section 2 to (49), with $\mu_1 = \tfrac{1}{2}\mu^2$,

$$\mathbf{u} = \begin{pmatrix} \eta \\ V \end{pmatrix}, \qquad \mathbf{D}_1 = \begin{pmatrix} \tfrac{1}{3}V_{\xi\xi\xi} \\ V_{\xi\xi\tau} \end{pmatrix},$$

$$\mathbf{C}_0 = \begin{pmatrix} 0 & 1 \\ 1 & 0 \end{pmatrix}, \qquad C_1 = \begin{pmatrix} V & \eta \\ 0 & V \end{pmatrix}.$$

The eigenvalues of $\mathbf{C}_0$ are $\lambda = \pm 1$. We choose $\lambda = 1$ and follow the right-running waves by introducing

$$X = \xi - \tau.$$

This leads to

$$\mathbf{u} = \begin{pmatrix} 1 \\ 1 \end{pmatrix} U,$$

where the scalar function U must satisfy the solvability conditions

$$U_{\tau_1} = -\tfrac{3}{2} U U_X \tag{50a}$$

and

$$U_{\tau_2} = -\tfrac{1}{3} U_{XXX}. \tag{50b}$$

Upon "reconstituting" the Korteweg-deVries equation, we get

$$U_\tau + \frac{3}{2}\epsilon U U_X + \frac{\mu^2}{6} U_{XXX} = 0. \tag{51}$$

If $9\epsilon \ll \mu^2$, then the time evolution of U is governed by (50b), with changes on the slow time τ_2 dominant. Equation (50b) is linear and dispersive; U therefore decays in time as the disturbance disperses. Consequently, the nonlinear term never amounts to more than a small perturbation in (51), which is therefore well approximated by (50b) for all time.

On the other hand, if $9\epsilon \gg \mu^2$, one cannot generally regard one member of the pair (50) as giving a good approximation for all time. In this case, the time evolution of U initially proceeds according to (50a), with the slow time τ_1 dominant. Depending upon the "initial" distribution of U regarded as a function of (X, τ_1), U will either relax through a family of simple waves or will steepen and tend to become multivalued. In the latter event, which is the case for any initial distribution of U that tends to zero as $|X| \to \infty$, the solution so obtained holds only for a finite time. Clearly, as the wave steepens the derivatives with respect to X become large, and ultimately $\mu^2 U_{\tau_2}/\epsilon$ must become comparable to U_{τ_1}. At such times, one must transfer attention to the composite equation (51). If multivaluedness of the solutions to (50a) is to be resolved by introducing shock waves as in Section 2, the appropriate conservation laws linking the two sides of the discontinuity must be provided by (51). That is, one must regard (50a) as embedded in the more general equation (51).

The question then is: Does the embedding of (50a) in (51) allow shocks to form? To examine this question, we rewrite (51) by putting

$$T = \tfrac{3}{2}\epsilon\tau, \qquad \delta = -\frac{1}{9}\frac{\mu^2}{\epsilon}$$

to obtain

$$U_T + UU_X = \delta U_{XXX}. \tag{52}$$

For the case of interest $|\delta| \ll 1$, and we wish to know if solutions of (52) with $\delta \neq 0$ tend to the shock solutions required by the limit equation ($\delta = 0$) as $\delta \to 0$. We shall address this question in Subsection 3d.

3c. Stationary Wave Solutions of the Korteweg-deVries Equation

Stationary (or permanent) waves are waves that propagate without change of form at a constant speed. Such waves are found by putting

$$\tilde{X} = X - \mathscr{S}T,$$

where $\mathscr{S}$ is the unknown wavespeed that must be determined. When substituted into the Korteweg-deVries equation (52), this yields the ordinary differential equation

$$\delta U''' = (U - \mathscr{S})U' \tag{53}$$

where the prime indicates differentiation with respect to $\tilde{X}$.

An integration yields

$$\delta U'' = \tfrac{1}{2}U^2 - \mathscr{S}U + \tfrac{1}{2}c_1,$$

where c_1 is an arbitrary constant of integration. Multiplication of the equation by U' followed by a second integration produces

$$(U')^2 = \delta^{-1}[\tfrac{1}{3}U^3 - \mathscr{S}U^2 + c_1U + c_2]. \tag{54}$$

Real solutions exist only for U such that

$$\delta^{-1}P(U) = \delta^{-1}[\tfrac{1}{3}U^3 - \mathscr{S}U^2 + c_1U + c_2] \geq 0.$$

The polynomial $P(U)$ may have one or three real roots. There are no bounded solutions in the event of a single real root. We assume, therefore, that P has three real roots $U_1 \leq U_2 \leq U_3$ and $P = \tfrac{1}{3}(U - U_1)(U - U_2) \cdot (U - U_3)$. The polynomial $P(U)$ is plotted in Figure IV.3 for three possibilities: In case A, all roots are distinct; in case B, roots U_2 and U_3 are coincident; and in case C, U_1 and U_2 are coincident.

Assuming $\delta > 0$ (the modifications required if δ is negative, as in the case of water waves, will be indicated later), solutions are possible for those portions of the curves A, B, C, for which P is nonnegative. Double

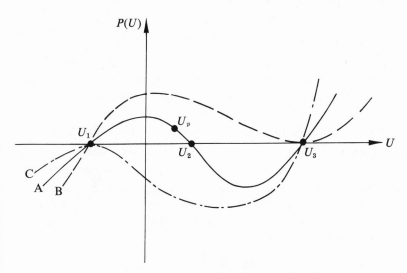

IV.3. The cubic $P(U)$ that determines the character of stationary waves.

roots of P in the figure correspond to values of U at either "initial" or "terminal" points of an integral curve of (54). At such points $U' = 0$, $U'' = 0$, and from (53), therefore, all derivatives of U vanish. Consequently, the constant value of U at a double root is an exact integral of (53). It follows that a nonconstant solution tends towards a double root only asymptotically as $\tilde{X} \to \infty$ or $\tilde{X} \to -\infty$.

Consider a point, say U_p, near a single zero of a curve such as A, and suppose $U'(U_p) > 0$, (see Figure IV.3). Since $U' > 0$ in (U_p, U_2), U will increase as $\tilde{X}$ increases, but since P has a finite negative slope in this portion of the curve, $U'' < 0$ and so U' must decrease. At U_2, U' has decreased to zero, but U'' remains negative. Therefore U' must become negative with an increase in $\tilde{X}$ and the solution "reflects" from this point, retracing its path on curve A. Thus curve A, with a positive portion sitting between two points where P vanishes with nonvanishing slope, represents a solution for $U(\tilde{X})$ that oscillates between U_1 and U_2.

The oscillation may be found explicitly in terms of the Jacobian elliptic function cn as

$$U = U_2 - (U_2 - U_1)cn^2\left[\sqrt{\frac{U_3 - U_1}{12\delta}}\ \tilde{X} \,\middle|\, m\right], \qquad (55)$$

where the parameter

$$m = \frac{U_2 - U_1}{U_3 - U_1}.$$

The function $cn(u \,|\, m)$ is defined in Abramowitz and Stegun (1964, secs. 6.1 and 7.2): $cn^2(u \,|\, m)$ oscillates between zero and unity with a period of $2K(m)$, where $K(m)$ is the complete elliptic integral of the first kind (Abramowitz and Stegun, sec. 17.3).

The infinite wavetrain (55) is called a "cnoidal" wave, and has a wavelength $2\sqrt{12\delta/(U_3 - U_1)}\ K(m)$. For the special values of $m = 0$ and $m = 1$, the function $cn(u \,|\, m)$, and hence (55), is described by more familiar functions, viz., $cn(u \,|\, 0) = \cos u$, and $cn(u \,|\, 1) = \operatorname{sech} u$. The amplitude $\tilde{a}$ of the oscillation about the mean value $\overline{U} = (U_1 + U_2)/2$ is $\tilde{a} = (U_2 - U_1)/2$. For waves of small amplitude, sinusoidal oscillations are recovered, since as $m \to 0$

$$U \sim \overline{U} - \tilde{a}\cos\left(2\tilde{X}\ \sqrt{\frac{U_3 - U_1}{12\delta}}\right).$$

Returning to case B of Figure IV.3, a solution approaching U_3 as $\tilde{X} \to -\infty$ is possible. It decreases to U_1 as the solution traces the curve B from U_3 to U_1, where it undergoes a "reflection," returning along B to U_3 as $\tilde{X} \to \infty$. Plotted against $\tilde{X}$, the solution looks like a single, symmetrical valley (or if $\delta < 0$, like a peak). This wave is therefore of infinite length, and is called a "solitary" wave. One can easily show that the solitary wave is given by

$$U = U_3 - (U_3 - U_1)\,\text{sech}^2\left(\tilde{X}\,\sqrt{\frac{a}{12\delta}}\right),$$

since the integration of (54) is elementary when $U_2 = U_3$. The result can also be found from (55) by letting U_2 approach U_3 so that $m \to 1$, and, therefore, $cn(u \,|\, m) \to \text{sech}\,u$. Since $K(m) \to \infty$ as $m \to 1$, the wavelength of the cnoidal waves tends to infinity in this solitary wave limit, as already noted. Here $a\ (= U_3 - U_1)$ is the amplitude of the solitary wave peak (or dip) above (or below) U_0.

The occurrence of cnoidal and solitary waves is probably best established in shallow water. A comprehensive account of their theory in this connection is given by Benjamin and Lighthill (1954), where the physical significance of these waves is elucidated.

Case C remains to be discussed: it has sometimes been said to lead to the hydraulic jump. A hydraulic jump, the water-wave analog of a shock wave, is a discontinuous disturbance consisting of two joined semi-infinite regions in which U takes different piecewise constant values. Referring to Figure IV.3, case C, the two conjugate values of U would be U_1 and U_3. The latter value is not a possible solution, however, since $U'' \neq 0$ there. The only solution in case C is the single constant value $U = U_1$. Actually, even if (53) did allow discontinuous solutions, the polynomial $P(U)$ would probably be of little use in their characterization, since it follows after two integrations in which no allowances for the inevitable discontinuities in U' and U'' are made.

We have, therefore, only two nonconstant waves of permanent form, the cnoidal waves of case A and the solitary wave of case B. The wave speed $\mathscr{S}$ is found from the roots U_1, U_2, U_3 by the expression

$$\mathscr{S} = \tfrac{1}{3}(U_1 + U_2 + U_3)$$

(since the coefficient of the quadratic term in a cubic is the negative of the sum of the roots).

Substituting for $\tilde{X}$, and generalizing for both positive and negative values of δ the final formulae for the cnoidal and solitary waves,

$$U = \overline{U} + \frac{\delta}{|\delta|} a_1 \left\{ 1 - cn^2 \left[\sqrt{\frac{a_1}{6m|\delta|}} \left(x - [\overline{U} + \frac{\delta}{|\delta|} \frac{2-m}{3m} a_1]T \right) | m \right] \right\} \quad (56)$$

and

$$U = U_0 - \frac{a\delta}{|\delta|} sech^2 \left[\sqrt{\frac{a}{12|\delta|}} \left(X + \left[\frac{1}{3} a \frac{\delta}{|\delta|} - U_0 \right] T \right) \right], \quad (57)$$

respectively, are found. In (56) the three independent parameters U_1, U_2, and U_3 that have so far characterized the cnoidal waves are replaced by the three equivalent adjustable parameters: $\overline{U}$ (the mean level of the oscillation), a_1 (the mean-to-peak and mean-to-trough amplitude), and m (taking values from zero to unity). Similarly, in the solitary wave solution (57), the two distinct parameters U_1 and U_3 are replaced by the undisturbed value, now indicated by U_0.

Expressions (56) and (57) show that both the wave speeds and wave shapes depend upon the amplitude of the disturbance and upon the background state upon which the wave propagates. The appearance of the amplitude is a feature due to nonlinearity. The dependence of the speed upon amplitude has been called "amplitude dispersion" and may be contrasted with the strictly linear dispersion due to wavelength encountered in Chapters I and II. Of course, it must be remembered that these nonlinear waves cannot generally be superposed.

3d. Does the Korteweg-deVries Equation Permit a
Shock Structure as $\delta \to 0$?

We have seen from Chapter III and Section 2 of this chapter that the equation

$$U_T + UU_X = 0 \quad (58)$$

has weak solutions with discontinuities. Thus such shock solutions may be the limit of smooth solutions of a more general equation [e.g., the Burgers equation (31)] as a small parameter tends to zero. Thus, by appropriate use of conservation laws, the solutions of the limit equation

(58) are the limit of the solutions of, for example, the Burgers equation. Furthermore, the discontinuity in solutions of (58) can be replaced by smooth transition such as (34).

Equation (58) is also the limit equation as the dispersive parameter $\delta \to 0$ in the K-dV equation. Because solutions of the limit equation must contain shocks if they are to make sense beyond the time that characteristics first intersect, it is of interest to know whether such solutions are the limit of solutions of the K-dV equation. That is, does $\lim_{\delta \to 0} U(X, T; \delta) = U(X, T; 0)$? A related question is whether the K-dV equation can replace a shock discontinuity by a smooth transition. Heuristic considerations that will be advanced here suggest that the solutions of the K-dV equation do not approach shocklike solutions in the usual sense as $\delta \to 0$. The behavior of these solutions is not well understood in this limit, but one may refer to Chapter VIII, p. 233, for a discussion of some of the possibilities. A model equation investigated by Yajima et al. (1966) is also of interest in this connection.

Conservation properties are the basis for weak solutions. If we consider an initial condition $U(X, 0) = U_0(X)$ such that the integrals

$$M_0 = \int_{-\infty}^{\infty} U_0(X)\, dX$$

and

$$E_0 = \int_{-\infty}^{\infty} \tfrac{1}{2} U_0^2(X)\, dX$$

are finite, then by integrating (52) one easily sees that, like the Burgers equation, the K-dV equation conserves "momentum," since

$$\frac{d}{dT} \int_{-\infty}^{\infty} U(X, T)\, dX = 0 \quad \text{or} \quad M \equiv \int_{-\infty}^{\infty} U(X, T)\, dX = M_0.$$

In contrast to the Burgers equation, however, by multiplying (52) by U and integrating, we see that the K-dV equation conserves the "energy" E:

$$E \equiv \int_{-\infty}^{\infty} \tfrac{1}{2} U^2(X, T)\, dX = E_0.$$

The K-dV equation is peculiar, as it actually is associated with an infinite number of conservation laws (see Chapter VIII), but the first two such laws are sufficient for our purposes.

The limit equation (58) can be made to conserve momentum or energy but not both. In this regard it is related to the Burgers equation and differs from the K-dV equation. To see this, suppose a shock is required to solve (58). For simplicity. assume that only a single discontinuity is required, propagating with speed $\mathcal{S} = \dot{X}_s(T)$. By integrating (58) from $-\infty$ to ∞, and interchanging orders of integration [taking care to account for the discontinuity at $X_s(T)$], one finds

$$\frac{dM}{dT} + \mathcal{S}[U] = \tfrac{1}{2}[U^2],$$

where $[f] = f(X_s +) - f(X_s -)$ is the jump of f across the shock. A similar treatment of the equation resulting from multiplying (58) by U shows that

$$\frac{dE}{dT} + \mathcal{S}[\tfrac{1}{2}U^2] = \tfrac{1}{3}[U^3].$$

In physical problems (gas dynamics for example), M usually corresponds to momentum, which is conserved, while E is related to kinetic energy, which need not be conserved. Putting $dM/dT = 0$ fixes the shock speed to be (in the present case)

$$\mathcal{S} = \dot{X}_s = \frac{1}{2}\frac{[U^2]}{[U]}.$$

Substituting into the energy equation, one finds that the rate of loss of energy is

$$\frac{dE}{dT} = \frac{1}{12}[U]^3,$$

which is necessarily negative.

The fact that $dE/dT = 0$ for all $\delta \neq 0$ but has a finite negative value for $\delta = 0$ is evidence that the limit solution for $\delta \to 0$ does not tend to solutions of the limit equation and, in particular, does not involve shock waves.

As a second approach (see Gel'fand, 1963), one may attempt to replace the discontinuity with a smooth structure supplied by the K-dV equation. We therefore assume the existence of a smooth transition, with a thickness that tends to zero with δ, connecting two different states. The thickness of this transition region will be denoted by $\Delta(\delta)$, $\Delta \to 0$ as $\delta \to 0$. We transform from (X, T) coordinates to stretched coordinates moving with the shock by introducing the coordinate

$$\zeta = \frac{X - X_s(T)}{\Delta},$$

where $X_s(T)$ is the (unknown) shock trajectory. If we now regard U to be a function of (ζ, T), the K-dV equation (multiplied by Δ) becomes

$$\delta\Delta^{-2}U_{\zeta\zeta\zeta} - (U - \mathscr{S})U_\zeta = \Delta U_T.$$

As $\delta \to 0$, the righthand side vanishes. The only nontrivial choice of Δ is $\Delta = O(\delta^{1/2})$, and without loss of generality we put $\Delta = \delta^{1/2}$ and pass to the limit $\delta \to 0$. The coordinate T now appears only as a parameter in the function $\mathscr{S}(T)$, and the limiting equation is

$$U_{\zeta\zeta\zeta} - (U - \mathscr{S})U_\zeta = 0.$$

This is simply the stationary K-dV equation (53). Boundary conditions in the stretched coordinates are $U \to U_1$ as $\zeta \to \infty$ and $U \to U_2$ as $\zeta \to -\infty$. The only solutions of (53), however, are the cnoidal waves and the solitary waves, and neither solution can fit the boundary conditions required of the shock transition. Therefore, the K-dV equation is not capable of supplying a shock structure, and the solutions of the limit equation (58) are not properly embedded in solutions of the K-dV equation.

3e. The Combined Korteweg-deVries and Burgers equation

Several variants of the K-dV equation have arisen in physical problems. We shall only mention one of these, the combined KortewegdeVries and Burgers equation:

$$U_T + UU_X = \delta_1 U_{XX} + \delta_2 U_{XXX}. \tag{59}$$

Equation (59) has been derived by Mei (1966) and Johnson (1970) for different physical situations. Stationary solutions are conveniently discussed in the phase plane (after an integration in X), but the final solution can no longer be given by quadratures. These solutions are thoroughly discussed in the review by Jeffrey and Kakutani (1972); much additional information concerning the Korteweg-deVries and other dispersive equations may also be found there.

CHAPTER V

Dispersive Waves and Variational Principles

Gerald B. Whitham

There are two main classes of wave motion. The first might be called "hyperbolic," since the governing partial differential equations are hyperbolic, and the second might be called "dispersive," since in the simplest examples a local disturbance disperses into a wavetrain. In general, dispersive waves are not governed by hyperbolic equations, and the elucidation of their mathematical structure and typical properties is a more complicated affair. The distinction between the two classes of waves is not always sufficiently well recognized. Students sometimes are left with the impression that the mathematical theory of hyperbolic equations is a sophisticated approach to wave motion and covers all cases, whereas the use of sinusoidal solutions

$$e^{i(kx-\omega t)}$$

with or without Fourier superposition, is an elementary approach to the same thing. The confusion is possible because both approaches may be used in an attack on the simple wave equation. However, the two approaches actually focus on the two different classes of wave motion. Furthermore, while many wave problems are hyperbolic, one could probably safely claim that the majority are not. In the most familiar example of water waves (treated in Chapter IV), the governing equation is Laplace's equation with strange boundary conditions at the free surface.

This chapter covers the main topics of the lectures given at Cornell University in 1969. It was written originally as an article for Volume 7 of the Mathematical Association of America's *Studies in Mathematics* (1971). Permission to reprint it here is gratefully acknowledged.

The research was supported by the Office of Naval Research, U. S. Navy.

The theory of hyperbolic equations, with its attendant discussion of characteristics, shock formation in nonlinear problems, weak solutions, and so on is well established. Even for linear dispersive problems, the corresponding questions of the general form of the equations and the general features of their solutions are much more complicated. For nonlinear dispersive problems such results were almost nonexistent. In the last few years, however, our understanding of dispersive waves, both linear and nonlinear, has advanced considerably. It has led to partial unification of points of view and techniques. This chapter is a review of some of these developments with emphasis on those that contribute to the unification and allow extension to nonlinear problems. A mathematical novelty is the description of waves by variational principles.

1. Dispersive Waves

The study of dispersive waves starts with periodic traveling wave-trains. In linear problems, these are sinusoidal and may be taken in the complex form

$$u(x, t) = ae^{i(kx - \omega t)}, \tag{1}$$

where a, k, ω are real constants. Here, a is the amplitude, k is the wavenumber (number of oscillations per 2π in space), and ω is the frequency (number of oscillations per 2π in time). The velocity of the wavetrain is the phase velocity $c = \omega/k$. The frequency ω and wavenumber k are related by a "dispersion relation"

$$G(\omega, k) = 0, \tag{2}$$

where the function G is determined by the particular equations of the problem. Some examples are the beam equation:

$$u_{tt} + \gamma^2 u_{xxxx} \equiv 0,$$

$$G(\omega, k) \equiv \omega^2 - \gamma^2 k^4 = 0, \tag{3}$$

and the linear Korteweg-deVries equation:

$$u_t + \alpha u_x + \beta u_{xxx} = 0,$$

$$G(\omega, k) \equiv \omega - \alpha k + \beta k^3 = 0. \tag{4}$$

Neither of these equations is hyperbolic. Indeed, integral equations may also admit wavetrain solutions. The equations

$$u_t + \int_{-\infty}^{\infty} K(x-\xi)u_\xi(\xi, t)\, d\xi = 0, \tag{5}$$

$$G(\omega, k) \equiv \omega - k \int_{-\infty}^{\infty} K(\zeta)e^{-ik\zeta}\, d\zeta = 0, \tag{6}$$

and

$$u_{tt} - \int_{-\infty}^{\infty} K_1(x-\xi)u_{\xi\xi}(\xi, t)\, d\xi = 0, \tag{7}$$

$$G(\omega, k) \equiv \omega^2 - k^2 \int_{-\infty}^{\infty} K_1(\zeta)e^{-ik\zeta}\, d\zeta = 0, \tag{8}$$

are examples. It should be noted in (6) that the phase velocity $c(k)$ is the Fourier transform of the kernel $K(x)$, and conversely the kernel $K(x)$ is the Fourier transform

$$K(x) = \frac{1}{2\pi} \int_{-\infty}^{\infty} c(k)e^{ikx}\, dk. \tag{9}$$

Similarly, the kernel $K_1(x)$ in (7) is the transform of $c^2(k)$. These are useful in proposing equations that will provide a *given* dispersion relation; the appropriate kernels K or K_1 are determined from the given $c(k)$ and used in (5) or (7). When $c(k)$ is a polynomial, as in (3) and (4), the kernels K, K_1 consist of δ-functions, and (5) and (7) reduce to the corresponding differential equations.

In the linear theory of water waves there are elementary solutions that lead to the expression (1) for the height of the surface, with the dispersion relation

$$G(\omega, k) \equiv \omega^2 - gk \tanh kh = 0, \tag{10}$$

where h is the undisturbed depth and g is the acceleration due to gravity. The analysis involves the vertical coordinate y as well as x and t, but the dependence on y is not wavelike. The linear problem is formulated in Chapter IV, Section 3.

This variety in the equations governing the wave motion shows that the unifying feature is the dispersion relation itself rather than the type of equation. For these linear problems one might almost view the equations as a mere source of the appropriate dispersion relation and push them into a subsidiary role. As noted above, an equation can always be constructed to match a dispersion relation. For simple linear partial differential equations of the form

$$P\left(\frac{\partial}{\partial t}, \frac{\partial}{\partial x}\right) u = 0,$$

as in (3) and (4), the dispersion relation is simply

$$P(-i\omega, ik) = 0,$$

with an immediate correspondence

$$\frac{\partial}{\partial t} \leftrightarrow -i\omega, \qquad \frac{\partial}{\partial x} \leftrightarrow ik,$$

between the two. It is clear that if such differential equations have a polynomial for P, they will always yield polynomial dispersion relations. Equation (10) is transcendental because of the non-wavelike dependence on y. The formulation of (5) from (9) and (10) has been pursued to some extent by Whitham (1967b) and Seliger (1968).

The extension to more space dimensions is immediate with a vector wavenumber $\mathbf{k}$ in (1) and (2) to give

$$u = ae^{i(\mathbf{k}\cdot\mathbf{x}-\omega t)}, \qquad G(\omega, \mathbf{k}) = 0. \tag{11}$$

The phase velocity has magnitude $c = \omega/|\mathbf{k}|$ and direction $\mathbf{k}$.

For linear problems, more general solutions are constructed by Fourier superposition of solutions of (1) or (11) for different wavenumbers $\mathbf{k}$ with corresponding frequencies $\omega(\mathbf{k})$ to satisfy the dispersion relation. The term "dispersion" refers to those cases for which the phase velocity is not the same for all values of $\mathbf{k}$. For in such cases, the Fourier components making up any given distribution will travel at different speeds and the disturbance will "disperse." It is convenient to express the condition for dispersive waves in terms of the roots $\omega(\mathbf{k})$ of the dispersion relation and to change the requirement slightly. We shall refer to linear waves as dispersive if

$$\text{determinant} \left|\frac{\partial^2 \omega}{\partial k_i \partial k_j}\right| \neq 0. \tag{12}$$

For one-dimensional waves this reduces to

$$\omega''(k) \neq 0. \tag{13}$$

Of course, these quantities might vanish for isolated values of $\mathbf{k}$ or in the limits of $\mathbf{k} \to 0$ and $\mathbf{k} \to \infty$. Equation (13) is almost equivalent to $c'(k) \neq 0$; it excludes $\omega = \alpha k + \beta$, whereas the latter does not. This special example should be excluded, however, because there is no real dispersion. It should also be noted that the simple wave equation with $\omega = \pm \alpha k$ has been eliminated from dispersive waves. The approach via (1) and Fourier superposition gives the correct results, but there is no dispersion. The behavior of the genuine cases with $\omega''(k) \neq 0$ is given in the next section.

The dispersion relation, its use in Fourier synthesis, and, as we shall see later, its use in direct asymptotic techniques allows a general treatment of linear problems. But this is clearly insufficient for nonlinear problems. We can identify dispersive wave problems by the existence of periodic wavetrains analogous to (1), but Fourier superposition cannot be used for further developments.

Probably the first nonlinear dispersive waves were discussed by Stokes in 1847 in his investigations of water waves. He found solutions to the nonlinear problem of water waves by expanding all quantities in powers of the amplitude, carefully keeping the expansions uniformly valid by allowing the frequency to depend on the amplitude. (The latter technique subsequently became known as Poincaré's method!) For deep water, the expansions for the surface elevation and frequency are

$$\eta = a \cos (kx - \omega t) + \tfrac{1}{2}ka^2 \cos 2(kx - \omega t) + \tfrac{3}{8}k^2a^3 \cos 3(kx - \omega t) + \cdots,$$
$$\omega^2 = gk + gk^3a^2 + \cdots. \tag{14}$$

(In this expression η is ϵh times the η used in Chapter IV.) Stokes' investigation introduces the key idea that the dispersion relation includes the amplitude; the phase velocity depends on both wavenumber and amplitude. Simpler examples in which the results corresponding to (14) are expressed in closed form appeared later. One of the most important ones, again from water waves, is the Korteweg-deVries equation (see Chapters IV and VIII), which is an approximation for water waves in shallow water. Its advantage is the elimination of the vertical coordinate, thereby resulting in a single equation for the surface elevation. It is

$$\eta_t + (c_0 + \alpha\eta)\eta_x + \beta\eta_{xxx} = 0, \tag{15}$$

where $c_0 = \sqrt{gh}$, $\alpha = 3c_0/2h$, $\beta = c_0h^2/6$. Uniform wavetrains are obtained from solutions

$$\eta = \eta(\theta), \qquad \theta = kx - \omega t. \tag{16}$$

When this form is substituted in (15), the resulting ordinary differential equation can be integrated twice to

$$\frac{1}{2}\beta k^2\left(\frac{d\eta}{d\theta}\right)^2 = A + B\eta + \frac{1}{2}\left(\frac{\omega}{k} - c_0\right)\eta^2 - \frac{1}{6}\alpha\eta^3, \tag{17}$$

where A, B are constants of integration. The solution may then be written in terms of the Jacobian elliptic function $cn\theta$, and, as a consequence, Korteweg and deVries named these waves "cnoidal waves." A limiting case [when the cubic on the right of (17) has a double zero] gives the famous solitary wave. (Details are given in Chapter IV, Subsection 3c.) In the periodic case, the relations of the amplitude and modulus of the elliptic function to the parameters ω, k provide the dispersion relation

$$G(\omega, k, a) = 0. \tag{18}$$

The general conclusion is that nonlinear dispersive waves are recognized by the existence of periodic wavetrains

$$u = U(\theta), \qquad \theta = kx - \omega t, \tag{19}$$

where U is a periodic function of the phase θ; the solution for U will bring in a parameter a, which is the amplitude, and the solution requires a dispersion relation in the form (18). As it turns out, there may be other important parameters that appear both in U and G.

A simpler example for some purposes, which has acquired recent interest, is a nonlinear version of the Klein-Gordon equation:

$$u_{tt} - u_{xx} + V'(u) = 0. \tag{20}$$

This happens to be hyperbolic, but we shall be concerned with dispersive behavior away from wavefronts. (The appearance of both hyperbolic and dispersive behavior in the same equation shows yet again the complications in the classification problem.) The periodic solution satisfies

$$(\omega^2 - k^2)U_{\theta\theta} + V'(U) = 0, \tag{21}$$

which has the first integral

$$\tfrac{1}{2}(\omega^2 - k^2)U_\theta^2 + V(U) = A. \tag{22}$$

The constant A can be used as a parameter equivalent to the amplitude a. Equation (25) can be solved to give the function $U(\theta)$ in the inverse form

$$\theta = \sqrt{\tfrac{1}{2}(\omega^2 - k^2)} \int \frac{dU}{\sqrt{A - V(U)}}. \tag{23}$$

We may normalize the period in θ to be 2π so that k and ω still determine the number of oscillations per 2π units of length and time. Then, it follows that

$$2\pi = \sqrt{\tfrac{1}{2}(\omega^2 - k^2)} \oint \frac{dU}{\sqrt{A - V(U)}}, \tag{24}$$

where $\oint$ denotes integration over a complete period. This is the dispersion relation between ω, k and A.

The next step is to see how more general solutions can be developed from these periodic wavetrains. In linear problems the Fourier synthesis is clear; we take this up next, before returning to the nonlinear problems.

2. Fourier Synthesis and Asympotic Behavior

Formally, a general solution

$$u = \int_{-\infty}^{\infty} F(k)e^{i\{kx - W(k)t\}}\, dk \tag{25}$$

is deduced from (1) and (2), where $\omega = W(k)$ is a specific solution of the dispersion relation (2). The complete solution will be the sum of terms like (25) with one integral for each of the solutions $\omega = W(k)$ of (2). The arbitrary functions $F(k)$ would be determined from appropriate initial or boundary conditions. These Fourier integrals give exact solutions, but their content is difficult to see and we are seeking ways to avoid them for later extension to nonlinear problems!

One starts to understand the main features of dispersive waves by considering the asymptotic behavior for large x and t. We recapitulate here some of the results of Chapter II. For wave motions, we are interested in traveling disturbances, which means that we are concerned with dependence on $x - ct$ for various c. Accordingly, we consider the asymptotic behavior of (25) as $t \to \infty$, with x/t held fixed. The solution (25) is written

$$u = \int_{-\infty}^{\infty} F(k)e^{-i\chi t}\, dk, \tag{26}$$

where

$$\chi(k) = W(k) - k\frac{x}{t}. \tag{27}$$

For the present, x/t is a fixed parameter, and only the dependence on k is displayed in χ. The integral in (26) may then be studied by the method of stationary phase; this is, in fact, the problem for which Kelvin developed the method. Kelvin argued that for large t, the main contribution to the integral is from the neighborhood of stationary points $k = k_0$ defined by

$$\chi'(k_0) = W'(k_0) - \frac{x}{t} = 0. \tag{28}$$

Otherwise the contributions oscillate rapidly and make little net contribution. The method and the related method of steepest descents, which is easier to use for the full asymptotic expansion, is now standard. The first term in the asymptotic behavior is

$$u(x, t) \sim F(k)\sqrt{\frac{2\pi}{t|\chi''(k)|}} \exp\left[-i\chi(k)t - i\frac{\pi}{4}\operatorname{sgn}\chi''(k)\right]$$

$$= F(k)\sqrt{\frac{2\pi}{t|W''(k)|}} \exp\left[ikx - iW(k)t - i\frac{\pi}{4}\operatorname{sgn}W''(k)\right]. \tag{29}$$

In this formula k is a function $k(x, t)$ determined by (28), and we have dropped the subscript zero.

The remarkable significance of (29) derives from the fact that it may be written just as in (1) as

$$u \sim ae^{i\theta} \tag{30}$$

with the important difference that a is no longer constant, nor are θ_x and θ_t. Indeed, with

$$\theta = kx - W(k)t,$$

we have, from (28),

$$\theta_x = k + \{x - W'(k)t\}k_x = k(x, t) \tag{31}$$

$$-\theta_t = W(k) - \{x - W'(k)t\}k_t = W(k). \tag{32}$$

Thus θ_x and $-\theta_t$ still have the significance of a wavenumber and a frequency, but they are no longer constant. At any point (x, t) they give the "local" wavenumber and "local" frequency. They are still related by the dispersion relation, and their variations with x and t are determined by (28). A vivid interpretation of (28) is provided by asking: how should an observer move if he wishes to follow a particular value k_1 of the local wavenumber? The answer is that he must keep

$$x = W'(k_1)t.$$

That is, he should move with the constant velocity $W'(k_1)$; this is the group velocity. We may embody this important result in the statement that values of the local wavenumber *propagate* with the group velocity. In contrast, to keep the same phase θ_1 such as a particular crest, an observer must travel with the phase velocity $W(k)/k$, where k is determined such that

$$\theta_1 = kx - W(k)t,$$

and this phase velocity is *not* constant. The distinction between group and phase velocities is crucial, and the former plays the dominant role.

The complex amplitude in (30) is

$$a = F(k)\sqrt{\frac{2\pi}{t|W''(k)|}} \exp\left[-i\frac{\pi}{4} \operatorname{sgn} W''(k)\right]. \tag{33}$$

The quantity $|a|^2$ is related to the energy density, and the significant form of (33) is related to the conservation of this quantity. Consider

$$Q(t) = \int_{x_1}^{x_2} |a|^2 \, dx$$

$$= 2\pi \int_{x_1}^{x_2} \frac{F(k)F^*(k)}{t |W''(k)|} \, dx. \tag{34}$$

In this integral, $k(x, t)$ is given by (28). Since k appears in the arguments of the integrand and x does not, it is natural to introduce k as a new variable of integration through the transformation

$$x = W'(k)t.$$

For $W''(k) > 0$, we have

$$Q(t) = 2\pi \int_{k_1}^{k_2} F(k)F^*(k) \, dk,$$

where k_1 and k_2 are defined by

$$x_1 = W'(k_1)t, \qquad x_2 = W'(k_2)t. \tag{35}$$

If $W''(k) < 0$, the limits are reversed. Now, if k_1 and k_2 are held fixed as t varies, $Q(t)$ remains constant. According to (35), the points x_1 and x_2 are then moving with the corresponding group velocities. We have shown, therefore, that the total amount of $|a|^2$ between points moving with the group velocity remains the same. *In this sense, "energy" propagates with the group velocity.* The points (35) separate with a distance proportional to t; hence, $|a|$ decreases like $t^{-1/2}$ The relation between $|a|^2$ and the true energy density will appear later.

It is striking that these important results depend only on the dispersion relation $\omega = W(k)$, and particularly on the expression $W'(k)$ for the group velocity. One then has the feeling that the Fourier analysis must be largely irrelevant, and one expects that similar results are feasible for nonlinear problems. We now turn to these questions.

The motivating step is to realize that in these asymptotic results we are dealing with "slowly varying" wavetrains. The asymptotic solution (30) is the same expression as in the uniform wavetrain (1), but the parameters a, $k = \theta_x$, $\omega = -\theta_t$ are no longer constant. They are, however, slowly varying in the sense that the relative change in one wavelength or one period is small. This is easily seen from (28) and (33). From (28),

$$\frac{k_t}{k} = -\frac{W'(k)}{kW''(k)}\frac{1}{t}, \qquad \frac{k_x}{k} = \frac{1}{kW''(k)}\frac{1}{t},$$

with similar results for ω, and from (33),

$$\frac{a_t}{a} = \left\{\frac{kF'(k)}{F(k)} - \frac{kW'''(k)}{2W''(k)}\right\}\frac{k_t}{k} - \frac{1}{2t},$$

with a similar expression for a_x/a; since the asymptotic expansion is for $t \to \infty$ and $x/t = W'(k)$ fixed, these are small.

With this motivation, we consider slowly varying wavetrains in a more general context. The idea is to start with the uniform wave-train, whether linear or nonlinear, extend the solution to allow the parameters (ω, k, a) and any others that arise to be slowly varying, and find direct ways of obtaining equations for these slowly varying parameters. By "slowly varying" we mean that their relative changes in one wavelength and in one period are small. The slow variations may come about in other ways than in the asymptotic behavior discussed above. It is of interest to take a signaling problem from a periodic source which is slowly modulated, or to consider propagation into a nonhomogeneous medium whose properties vary slowly in space or time. In these cases the small parameter ϵ for the slow variations will be provided in the boundary conditions or the equations, whereas it was a typical period divided by t itself in the case studied above.

We shall proceed intuitively at first in order to propose methods and derive results. The justification by formal perturbation expansions will be explained in Section 8.

3. Simple Derivation of Group Velocity Concepts for Linear Problems

We know from Section 2 that the group velocity plays two distinct roles. In one it determines the propagation of wavenumber and frequency; in the other, it determines variations in amplitude. In the asymptotic solution (30), one role is concerned with the determination of θ, the other with the determination of a.

To describe a slowly varying wavetrain, there must be a phase function $\theta(x, t)$. A local wavenumber $k(x, t)$ and frequency $\omega(x, t)$ are defined in terms of θ by

$$k(x, t) = \frac{\partial \theta}{\partial x}, \qquad \omega(x, t) = -\frac{\partial \theta}{\partial t}. \tag{36}$$

In view of the assumed slow variations of k, ω, it is reasonable to propose that these quantities still satisfy the dispersion relation. For linear problems the dispersion relation is

$$G(\omega, k) = 0. \tag{37}$$

Equations (36) and (37) give a nonlinear partial differential equation for θ. It is convenient to determine k and ω from (37) and the consistency relation [elimination of θ from (36)]

$$\frac{\partial k}{\partial t} + \frac{\partial \omega}{\partial x} = 0. \tag{38}$$

If $\omega = W(k)$ is a solution of (37), then (38) gives

$$\frac{\partial k}{\partial t} + C(k)\frac{\partial k}{\partial x} = 0, \qquad C(k) = W'(k). \tag{39}$$

This shows immediately that k is constant on characteristic curves defined by

$$\frac{dx}{dt} = C(k)$$

in the (x, t) plane. Since k is constant on each curve, the curves are straight lines, each with slope $C(k)$ corresponding to the value of k on that curve. The solution for an initial distribution $k = f(x)$ at $t = 0$ is

$$k = f(\xi), \qquad x = \xi + C(k)t. \tag{40}$$

In more physical terms, values of k propagate with the corresponding value of the group velocity $C(k)$. When the disturbance is concentrated initially at the origin, k may be determined from

$$x = C(k)t. \tag{41}$$

This is the case in (28); viewed from large x and t, the initial disturbance *is* effectively concentrated at the origin. In this case, but only in this case,

$$\theta = kx - W(k)t.$$

We thus have a very simple and yet more general derivation of the kinematics of the waves.

The second role of the group velocity is in the determination of a, and this involves dynamics. One looks for a differential equation for a to match the one for k in (39). It is not hard to find by various routes. Leaning first on (34) and (35), we note that

$$\frac{dQ}{dt} = \frac{d}{dt} \int_{x_1}^{x_2} |a|^2 dx = \int_{x_1}^{x_2} \frac{\partial}{\partial t} |a|^2 dx + |a|_2^2 C(k_2) - |a|_1^2 C(k_1) = 0.$$

As $x_2 - x_1 \to 0$, we have

$$\frac{\partial}{\partial t} |a|^2 + \frac{\partial}{\partial x} \{C(k)|a|^2\} = 0 \tag{42}$$

in the limit. Alternatively, it is well known in specific problems that the energy density and energy flux are proportional to $|a|^2$ and $C(k)|a|^2$, respectively. This is in fact another popular way to argue that energy propagates with the group velocity. Equation (42) is the corresponding equation for the conservation of energy. There are many variants of this argument that may be used to derive (42); but, until recently, they all suffered from the deficiency that (42) had to be established separately from the governing equations for each specific problem. Why is the answer always the same in terms of the group velocity? This has now been resolved by a general approach using variational principles. At the same time it is shown that (42) is not basically the equation for energy conservation, but rather the conservation of "wave action", which in simple cases is proportional to the energy.

For nonlinear problems, the introduction of θ and $k = \theta_x$, $\omega = -\theta_t$, is still sound. But the dispersion relation is now

$$G(\omega, k, a) = 0$$

and involves a. Thus, the two equations corresponding to (38) and (42) become coupled. The variational approach has also led to a simple derivation of the corresponding equations. Before we take this up in Section 5, the linear kinematics is extended in the next section to more dimensions, which could also be done by stationary phase on multiple Fourier integrals analogous to Section 2, and to nonhomogeneous media, which cannot be studied by Fourier integrals. A few examples are also given to show the power of these simple methods and to show the value of this type of more intuitive discussion even when the exact solution in terms of Fourier integrals is known.

4. Extensions and Examples

In more dimensions, the phase $\theta(\mathbf{x}, t)$ is a function of the vector position $\mathbf{x}$ and the time t. The vector wavenumber $\mathbf{k}$ and frequency ω are defined by

$$k_i = \frac{\partial \theta}{\partial x_i}, \qquad \omega = -\frac{\partial \theta}{\partial t}. \tag{43}$$

For linear problems in a homogeneous medium, the dispersion relation is $G(\omega, \mathbf{k}) = 0$. This relation will involve the parameters of the medium. For a nonhomogeneous medium, these parameters will be slowly varying functions of $\mathbf{x}$ or t, and the dispersion relation becomes

$$G(\omega, \mathbf{k}; t, \mathbf{x}) = 0. \tag{44}$$

We still propose to use it in conjunction with (43). The cross-elimination of θ from (43) gives

$$\frac{\partial k_i}{\partial t} + \frac{\partial \omega}{\partial x_i} = 0, \qquad \frac{\partial k_i}{\partial x_j} - \frac{\partial k_j}{\partial x_i} = 0. \tag{45}$$

If $\omega = W(\mathbf{k}; t, \mathbf{x})$ is taken as a solution of (44), the first equation in (45) gives

$$\frac{\partial k_i}{\partial t} + \frac{\partial W}{\partial k_j} \frac{\partial k_j}{\partial x_i} = -\frac{\partial W}{\partial x_i}, \tag{46}$$

where the summation convention has been used. We now introduce the vector group velocity defined by

$$C_j = \frac{\partial W}{\partial k_j} \tag{47}$$

and note from the second relation in (45) that $\partial k_j/\partial x_i$ may be replaced by $\partial k_i/\partial x_j$. Then

$$\frac{\partial k_i}{\partial t} + C_j \frac{\partial k_i}{\partial x_j} = -\frac{\partial W}{\partial x_i}. \tag{48}$$

In characteristic form, the equation may be written

$$\frac{dk_i}{dt} = -\frac{\partial W}{\partial x_i}, \qquad \frac{dx_i}{dt} = \frac{\partial W}{\partial k_i} = C_i. \tag{49}$$

These are Hamilton's equations with the k_i in the role of the generalized momenta and the frequency W in the role of the Hamiltonian. This is the well-known duality exploited in quantum theory. The equation for θ is

$$\frac{\partial \theta}{\partial t} + W\left(\frac{\partial \theta}{\partial x_i}, t, x_i\right) = 0;$$

this is the Hamilton-Jacobi equation.

In homogeneous media the wavenumber $\mathbf{k}$ still propagates unchanged with the group velocity $\mathbf{C}(\mathbf{k})$. In nonhomogeneous media, it propagates with this velocity, but the values vary at the rate $-\partial W/\partial \mathbf{x}$. In homogeneous media the general solution corresponding to (40) is

$$k_i = f_i(\xi),$$
$$x_i = \xi_i + C_i(\mathbf{k})t.$$

For a concentrated initial distribution, $\mathbf{k}$ is determined by

$$x_i = C_i(\mathbf{k})t. \tag{50}$$

[This would be the stationary point corresponding to (28) in the multiple Fourier analysis.]

A few typical and interesting examples are taken from the theory of water waves.

4a. Ocean Swell from Storms

In deep water, the dispersion relation (10) may be approximated by its limiting form $\omega = \sqrt{gk}$, for $kh \to \infty$. For two dimensions this becomes

$$\omega = W(\mathbf{k}) = \sqrt{gk}, \qquad k = |\mathbf{k}|. \tag{51}$$

For the wave distribution at large distances from a concentrated storm, it is appropriate to use (50). The pattern is symmetric, since $W(\mathbf{k})$ depends only on $|\mathbf{k}|$ and not on its direction. The group velocity is

$$C_i = \frac{1}{2}\sqrt{\frac{g}{k}}\frac{k_i}{k}, \qquad C = |\mathbf{C}| = \frac{1}{2}\sqrt{\frac{g}{k}}.$$

Hence, from (50),

$$r = |\mathbf{x}| = \frac{1}{2}\sqrt{\frac{g}{k}}\, t;$$

therefore,

$$k = \frac{1}{4}\frac{gt^2}{r^2}, \qquad \omega = \frac{1}{2}\frac{gt}{r}.$$

The formula has been applied to waves produced by storms in the South Pacific (Snodgrass et al., 1966). At distances of the order of 2000 miles, the frequency was found to vary linearly with t, and the constant of proportionality gave a very accurate determination of the distance of the storm.

4b. Ship Waves

The wave pattern produced by a ship (or other moving object) traveling with uniform velocity U in the negative x_1 direction on the surface of deep water is most easily treated as a steady flow problem relative to the ship. Then $\omega = 0$, but the two components (k_1, k_2) of the wavenumber are still related by a dispersion relation, and the geometry is determined according to (45) from

$$\frac{\partial k_1}{\partial x_2} - \frac{\partial k_2}{\partial x_1} = 0. \tag{52}$$

The dispersion relation is obtained by a transformation of (51) to a moving frame; it becomes

$$Uk_1 + \sqrt{gk} = 0, \qquad k = (k_1^2 + k_2^2)^{1/2}. \tag{53}$$

The characteristic form of (52) shows that the various values of k_1 and k_2 will be distributed on lines

$$\frac{dx_1}{dx_2} = -\frac{dk_2}{dk_1}, \tag{54}$$

where $k_2 = f(k_1)$ is determined from (53). It is easily verified that the righthand side of (54) has a minimum value of $2\sqrt{2}$ for all values of k_1. Hence, the wave pattern is contained in a wedge shaped region spreading out behind the ship, and the semi-angle of the wedge is $\cot^{-1} 2\sqrt{2} = 19.5°$ (Kelvin's famous result). Full details of the wave pattern can also be found; the version following this approach is given in Whitham (1961).

4c. Waves in a Stratified Fluid

In the so-called Boussinesq approximation, the dispersion relation is

$$\omega^2 = \frac{\omega_0^2 k_1^2}{(k_1^2 + k_2^2)},$$

where the density ρ_0 is stratified in the x_2 direction and

$$\omega_0^2 = -\frac{g}{\rho_0} \frac{d\rho_0}{dx_2}.$$

First, it is clear that waves are only possible if $\omega < \omega_0$. Second, the group velocity

$$\mathbf{C} = \left(\frac{\omega_0 k_2^2}{k^3}, -\frac{\omega_0 k_1 k_2}{k^3} \right)$$

is perpendicular to the wavenumber $\mathbf{k}$. Hence, the group velocity is perpendicular to the phase velocity; this is a striking example of the distinction between the two. Third, if waves are stimulated by a vibrating source with a fixed ω, an analysis similar to the ship-wave case shows that the waves are confined to a wedge-shaped region, making an angle $\sin^{-1} (\omega/\omega_0)$ with the horizontal. Full details and impressive photographs of experiments showing all these features are given in Mowbray and Rarity (1967).

5. Variational Principles

We return now to the general discussion and continue the development from Section 3. As noted there, the appropriate approach is through variational principles. The power of the method appears immediately, since it is just as easy to do nonlinear problems as linear ones! The nonlinear Klein-Gordon equation (20) will be used to develop the ideas. Ultimately, all that will be used is its corresponding Lagrangian, so the results can be taken over in general to any Lagrangian.

Equation (20) may be derived from the variational principle

$$\delta \int \int L(u_t, u_x, u)\, dt\, dx = 0, \tag{55}$$

where

$$L = \frac{1}{2}u_t^2 - \frac{1}{2}u_x^2 - V(u). \tag{56}$$

The periodic solution takes the form $u = U(\theta)$, where $\theta_x = k$ and $\theta_t = -\omega$ are constants, and the actual solution for $U(\theta)$ will bring in another constant A, which is equivalent to the amplitude. We now consider trial functions of this form. The Lagrangian becomes

$$L^{(0)} = L(-\omega U_\theta, kU_\theta, U). \tag{57}$$

Variations of this with respect to U must yield the equation for the periodic solution. For the Klein-Gordon example,

$$L^{(0)} = \frac{1}{2}(\omega^2 - k^2)U_\theta^2 - V(U), \tag{58}$$

and, indeed, (21) follows from its variational equation

$$\frac{d}{d\theta}\left(\frac{dL^{(0)}}{dU_\theta}\right) - \frac{dL^{(0)}}{\partial U} = 0. \tag{59}$$

But our aim is to go further and find equations for (ω, k, A) when they are allowed to vary. We now describe a procedure based on intuitive arguments; its formal justification will be provided later. For the uniform wavetrain, calculate the "average" Lagrangian

$$\mathscr{L}(\omega, k, A) = \frac{1}{2\pi}\int_0^{2\pi} L(-\omega U_\theta, kU_\theta, U)\, d\theta, \tag{60}$$

in terms of the constant parameters (ω, k, A). (There is some subtlety in doing this, to which we return shortly.) We propose that in the extension to slowly varying wavetrains, the slowly varying functions ω, k, A will satisfy the variational equations given by the "averaged" variational principle

$$\delta \int\int \mathscr{L}(\omega, k, A)\, dx\, dt = 0. \tag{61}$$

In this variational principle, ω and k cannot be varied independently; they are related by

$$\omega = -\theta_t, \qquad k = \theta_x.$$

The variational equations resulting from variations of δA and $\delta\theta$ in (61) are

$$\delta A: \qquad \mathscr{L}_A = 0 \tag{62}$$

and

$$\delta\theta: \quad \frac{\partial}{\partial t}\mathscr{L}_\omega - \frac{\partial}{\partial x}\mathscr{L}_k = 0, \tag{63}$$

respectively. Once (63) has been obtained, it is convenient to work with ω and k and complete the system with the consistency relation

$$\frac{\partial k}{\partial t} + \frac{\partial \omega}{\partial x} = 0. \tag{64}$$

This is the whole theory! Equation (62) is an equation relating ω, k, A, so it *must* be the dispersion relation. Equation (63) must be the sought-after differential equation for the amplitude. We now consider the content in detail.

5a. Linear Problems

For any linear problem, the Lagrangian must be quadratic in u and its derivatives. The periodic solution may be taken in the form

$$U = a \cos \theta. \tag{65}$$

Hence, when this is substituted in (60), the average Lagrangian *must* be proportional to a^2 and take the form

$$\mathscr{L} = G(\omega, k)a^2. \tag{66}$$

Then, (62) becomes

$$G(\omega, k) = 0. \tag{67}$$

Therefore, without detailed calculation, we know that the function appearing in (66) *must* be the linear dispersion function. It is interesting to note that the stationary value is $\mathscr{L} = 0$. In those cases where $\mathscr{L}$ is the difference of kinetic and potential energy, this shows the equipartition of energy between the two. It is also clear that the dispersion relation itself should not be used in substituting the periodic trial functions (65). This would lead to $\mathcal{L} = 0$; we would already have the stationary value, and there would not be enough freedom to apply the variational argument.

The amplitude equation (63) becomes

$$\frac{\partial}{\partial t}(G_\omega a^2) - \frac{\partial}{\partial x}(G_k a^2) = 0. \tag{68}$$

If a solution of (67) is $\omega = W(k)$, then $G(W, k) = 0$ and

$$G_\omega W'(k) + G_k = 0.$$

Hence, the group velocity C can be written in terms of G as

$$C = -\frac{G_k}{G_\omega}. \tag{69}$$

Thus, (68) takes the form

$$\frac{\partial}{\partial t}\left\{ g(k)a^2 \right\} + \frac{\partial}{\partial x}\left\{ g(k)C(k)a^2 \right\} = 0. \tag{70}$$

In this equation, the factor $g(k)$ can be slipped out, for it may be expanded to

$$g(k)\left\{ \frac{\partial(a^2)}{\partial t} + \frac{\partial}{\partial x}(Ca^2) \right\} + g'(k)\left\{ \frac{\partial k}{\partial t} + C(k)\frac{\partial k}{\partial x} \right\}a^2 = 0. \tag{71}$$

The last term vanishes from (64), so we have

$$\frac{\partial(a^2)}{\partial t} + \frac{\partial}{\partial x}\left\{ C(k)a^2 \right\} = 0. \tag{72}$$

This is the general proof of (42). The identification of (63) and its relation to the energy equation is left until we have discussed the nonlinear case.

5b. Klein-Gordon Example

In nonlinear problems the manipulations to obtain the function $\mathcal{L}$ are more subtle, and it is as well to do a specific example before the general case. For the nonlinear Klein-Gordon equation, the averaged Lagrangian is first obtained from (58) and (60) in the form

$$\mathcal{L} = \frac{1}{2\pi}\int_0^{2\pi} \left\{ \frac{1}{2}(\omega^2 - k^2)U_\theta^2 - V(U) \right\} d\theta. \tag{73}$$

The periodic solution satisfies (22). This is used to write (73), successively, as

$$\mathcal{L} = \frac{1}{2\pi}\int_0^{2\pi} (\omega^2 - k^2)U_\theta^2 \, d\theta - A$$

$$= \frac{1}{2\pi}\int_0^{2\pi} (\omega^2 - k^2)U_\theta \, dU - A$$

$$= \frac{1}{2\pi}\sqrt{2(\omega^2 - k^2)} \oint \sqrt{A - V(U)} \, dU - A, \tag{74}$$

where again $\oint$ denotes the integral over a complete period of the integrand. The final form does not require the dependence of U on θ. According to (62), the dispersion relation should be

$$\mathcal{L}_A = \frac{1}{2\pi}\sqrt{\frac{1}{2}(\omega^2 - k^2)} \oint \frac{dU}{\sqrt{A - V(U)}} - 1 = 0,$$

and, correctly, this checks with (24). In the linear limit, $V = \frac{1}{2}U^2$, and the dependence on A drops out. In fact, in this limit (74) becomes

$$\mathcal{L} = (\sqrt{\omega^2 - k^2} - 1)A;$$

since A is then proportional to a^2, this conforms to the general arguments.

5c. Hamiltonian Transformation

The transformation from (73) to (74) is not only a matter of convenience. The introduction of A into (73) uses information from the first integral (22) of the periodic solution, and yet the form of the trial function $U(\theta)$ must be left flexible enough to allow the variational argument to go through. For example, (22) implicitly contains the dispersion relation [as we saw in (24)], and we do not want to introduce that explicitly; it should be a consequence of the variational principle.

This is an important point that can be fully explained only by following the formal expansion procedure described in Section 8. Here, it is essential that the procedure should be unambiguous at least. The procedure is as follows.

The periodic solution is given by (59), which always has the first integral

$$U_\theta \frac{\partial L^{(0)}}{\partial U_\theta} - L^{(0)} = A. \tag{75}$$

Define a new variable

$$\Pi = \frac{\partial L^{(0)}}{\partial U_\theta} \tag{76}$$

in place of U_θ, and let

$$H(\Pi, U) = U_\theta \frac{\partial L^{(0)}}{\partial U_\theta} - L^{(0)}. \tag{77}$$

This is similar to the Hamiltonian transformation in mechanics. The integral (75) is then

$$H(\Pi, U) = A. \tag{78}$$

Now, from (75),

$$\mathscr{L} = \frac{1}{2\pi} \int_0^{2\pi} L^{(0)} \, d\theta$$

$$= \frac{1}{2\pi} \int_0^{2\pi} \{\Pi U_\theta - A\} \, d\theta$$

$$= \frac{1}{2\pi} \oint \Pi \, dU - A. \tag{79}$$

Finally, we can determine Π from (78) as a function of U and the parameters ω, k, A to write the loop integral as a function of ω, k, and A.

5d. Higher-Order Systems; Nonuniform Media

The theory goes through in a similar way when the Lagrangian in (55) contains more than one variable u and higher derivatives than the first. One feature is added with more dependent variables. Some of the additional variables appear only through their derivatives; they are potentials whose derivatives are the significant physical quantities. If ϕ denotes such a variable, the most general periodic solution for the physical quantities will require ϕ to be taken in the form

$$\phi = \beta x - \gamma t + \Phi(\theta),$$

where $\Phi(\theta)$ is periodic. The β and γ are extra paramenters in the solution. In the extension to slowly varying wavetrains, the quantity $\beta x - \gamma t$ has to be treated in the same way as $\theta = kx - \omega t$; in other words, we must take

$$\phi = \psi(x, t) + \Phi(\theta)$$

and define β, γ by

$$\beta = \psi_x, \qquad \gamma = -\psi_t.$$

These derivatives of the function ψ will appear in the averaged Lagrangian, and variations $\delta\psi$ will give the equation

$$\frac{\partial}{\partial t}\mathcal{L}_\gamma - \frac{\partial}{\partial x}\mathcal{L}_\beta = 0$$

in analogy to (63). Other constants of integration similar to A lead to equations like (62) and provide the appropriate additional relations between them (Whitham, 1965b, 1967b, 1970).

The extensions to more space variables are immediate. The wavenumber becomes a vector $k_i = \theta_{x_i}$ and the variation of (61) with respect to θ gives

$$\frac{\partial}{\partial t}\mathcal{L}_\omega - \frac{\partial}{\partial x_i}\mathcal{L}_{k_i} = 0$$

as the extension of (63).

For nonuniform media, the original Lagrangian in (55) will also have explicit dependence on $\mathbf{x}$ or t through the parameters that describe the medium. The periodic solution and the average Lagrangian $\mathcal{L}$ are calculated holding those parameters fixed. They are then freed in (61); that is,

$$\mathcal{L} = \mathcal{L}(\omega, \mathbf{k}, A; \mathbf{x}, t).$$

The extra dependence on $\mathbf{x}$ or t does not affect the variational argument, and (62) and (63) apply. Since the dispersion relation involves $\mathbf{x}$ or t, the coefficients in (68) will do also, and equations such as (72) will pick up additional terms.

The single inclusive form to cover all these various possibilities is one of the successes of the theory.

6. Adiabatic Invariants, Wave Action, and Energy

It is interesting to note—both for its own sake and as a guide to identifying equation (63)—that the present approach can be used in the corresponding problems of oscillating systems in dynamics and provides a different derivation of some of the standard results there. In dynamics, the governing equations are ordinary differential equations for functions of t, and the present theory can be applied by dropping the dependence on x. Periodic solutions can then be modulated only through a slowly varying parameter, $\lambda(t)$ say, and the Lagrangian in (55) takes the form $L[q_t, q, \lambda(t)]$. A nonlinear oscillator with $L = \frac{1}{2}q_t^2 - V(q, \lambda)$ would be an example closely related to (56).

The analysis proceeds exactly as described in Section 5, with the x dependence omitted at each step. The results are

$$\frac{d}{dt}\mathscr{L}_\omega = 0, \qquad \mathscr{L}_E = 0, \tag{80}$$

and

$$\mathscr{L} = \frac{\omega}{2\pi}\oint p\,dq - E, \tag{81}$$

where p is the usual generalized momentum, and $p(\lambda, E)$ is determined from the energy integral

$$H(p, q, \lambda) = E. \tag{82}$$

[In this case the quantity Π introduced in (76) reduces to $\Pi = \omega p$, and it is more convenient to use the usual momentum p.] The first result in (80) shows that

$$I(\lambda, E) = \mathscr{L}_\omega = \frac{1}{2\pi}\oint p\,dq \tag{83}$$

is constant for the slow variations; it is the so-called "adiabatic invariant." Its expression as the loop integral $\oint p\,dq$ is well known, but the expression as $\mathscr{L}_\omega$ is new. Of course, the energy E is not constant as the parameter λ varies. The second relation in (80) determines the frequency ω by

$$\omega \frac{\partial I}{\partial E} = 1,$$

which is also well known.

In the waves problem, the adiabatic equation becomes the conservation equation (63)

$$\frac{\partial}{\partial t}\mathcal{L}_\omega - \frac{\partial}{\partial x}\mathcal{L}_k = 0. \tag{84}$$

There is a nice duality between the quantities $\mathcal{L}_\omega$ and $\mathcal{L}_k$. If the wavetrain is uniform in x, but the medium changes in t, $\mathcal{L}_\omega = const.$; if a periodic wavetrain enters a region where the medium varies with x, $\mathcal{L}_k = const.$ In the propagation of modulations along a wavetrain, changes of $\mathcal{L}_\omega$ in time are balanced by changes of $\mathcal{L}_k$ in space. This is the conservation of something, and it is convenient to call it "wave action"; $\mathcal{L}_\omega$ and $-\mathcal{L}_k$ are the density and flux of wave action.

The application in dynamics has shown the crucial distinction between (84) and the energy equation. The distinction carries over to the case of waves. The energy equation, obtained from considerations of time-invariance in the variational principle, is

$$\frac{\partial}{\partial t}(\omega\mathcal{L}_\omega - \mathcal{L}) - \frac{\partial}{\partial x}(\omega\mathcal{L}_k) = -\mathcal{L}_t. \tag{85}$$

This may be established directly from (62)–(64), but the derivation from time-invariance identifies it as the energy equation. For a medium independent of time, the righthand side is zero and (85) is also in conservation form.

For linear problems, since $\mathcal{L}$ is quadratic in a, (85) also takes the form (70), but with a different function for g. For a uniform medium, g is a function of k and may be eliminated as before to give (72). Thus, (72) *is* related to the energy equation, but the wave action is both more fundamental and more convenient for nonhomogeneous media.

We may note one further point in linear problems. The stationary value is $\mathcal{L} = 0$, so the energy density

$$E = \omega\mathcal{L}_\omega.$$

Hence, the wave-action equation may be written

$$\frac{\partial}{\partial t}\left(\frac{E}{\omega}\right) + \frac{\partial}{\partial x}\left(\frac{CE}{\omega}\right) = 0. \tag{86}$$

The expression of the adiabatic invariant as E/ω for linear problems in mechanics is well known. Bretherton and Garrett (1969) claim to encounter some difficulties in applying these ideas to moving media, owing to mysteries in the transformation between moving frames, but no difficulties exist. In the reference frame moving with the medium, an observer will use $\mathscr{L}_0 = G_0(\omega_0, k)a^2$, say, while an observer who takes the medium to a velocity V will use $\mathscr{L} = G(\omega, k)a^2$, where the relation between them is $\omega = \omega_0 + Vk$ and $G(\omega, k) = G_0(\omega - Vk, k)$. In the first frame

$$E_0 = \omega_0 \mathscr{L}_{\omega_0} = \omega_0 G_{0\omega_0} a^2,$$

and in the second frame $E = \omega \mathscr{L}_\omega = \omega G_\omega a^2$. Since $G(\omega, k) = G_0(\omega - Vk, k)$, it follows that

$$C = -\frac{G_k}{G_\omega} = V - \frac{G_{0k}}{G_{0\omega_0}} = V + C_0$$

and

$$\frac{E}{\omega} = \frac{E_0}{\omega_0}.$$

Thus, (86) is correct, but it could more usefully be rewritten as

$$\frac{\partial}{\partial t}\left(\frac{E_0}{\omega_0}\right) + \frac{\partial}{\partial x}\left[(V + C_0)\frac{E_0}{\omega_0}\right] = 0.$$

We may also obtain the momentum equation by using the variational principle; it is derived from considerations of space invariance as

$$-\frac{\partial}{\partial t}(k\mathscr{L}_\omega) + \frac{\partial}{\partial x}(k\mathscr{L}_k - \mathscr{L}) = -\mathscr{L}_x,$$

which may also be verified directly from (62)–(64).

7. Nonlinear Group Velocity—Stability of Periodic Waves

For linear problems, the group velocity is the characteristic velocity for (64) and appears in the calculation of the distribution of $k(x, t)$. With k known, it appears a second time as the characteristic velocity in solving (63) [or (72)] for $a(x, t)$. For nonlinear problems, the equations (63) and (64) for k and a are coupled together. If these form a hyperbolic pair, it is natural to extend the definition of group velocity to the characteristic velocities of this pair. The two velocities are no longer equal for nonlinear problems. This will make important differences in the propagation of a modulation down the wavetrain. For a nonlinear wavetrain, an initially concentrated disturbance should eventually split into two disturbances. It would be of great interest to see whether this can be observed.

If the equations (63) and (64) are elliptic, it is easy to deduce that perturbations in x grow exponentially in time. This shows that the periodic wavetrain is itself unstable. Coincidentaly with the development of this theory, Benjamin (1967) deduced that deep-water waves were unstable in this sense. Noting that experimentally produced periodic wavetrains always showed instability after several wavelengths from the wave maker, he did an ingenious "sideband" instability analysis, using the second approximation beyond linear theory, to explain it. The method presented here was then used for waves of moderate amplitude in water of finite depth h (Whitham, 1967a). It was found that the equations were elliptic for $kh > 1.36$ and hyperbolic for $kh < 1.36$. Thus water waves of finite amplitude should be unstable for $kh > 1.36$.

The analysis for moderate amplitude is very simple when the only parameters are (ω, k, a); it becomes more involved when further parameters such as the β and γ referred to in Section 6 arise. Deep-water waves do involve only (ω, k, a), but the finite-depth case requires additional parameters that refer to small but important changes in mean depth and mean mass flow produced by the waves. The simpler case is presented here. In such problems, with moderately small amplitude, the main coupling between (63) and (64) is through the dependence of ω on a in the dispersion relation. If this is written to first order in a^2 as

$$\omega = \omega_0(k) + \omega_1(k)a^2, \tag{87}$$

it turns out to be sufficient to use

$$\frac{\partial k}{\partial t} + \frac{\partial}{\partial x}\{\omega_0(k) + \omega_1(k)a^2\} = 0 \tag{88}$$

and retain (72) in the linear approximation

$$\frac{\partial a^2}{\partial t} + \frac{\partial}{\partial x}\{C_0(k)a^2\} = 0, \qquad C_0 = \omega_0'(k). \tag{89}$$

A simple calculation shows that the characteristic velocities are

$$C = C_0 \pm a\sqrt{\omega_1 C_0'} + O(a^2). \tag{90}$$

If $\omega_1 C_0' > 0$, the double characteristic velocity C_0 splits into the two real roots given by (90), the system is hyperbolic, and (90) gives the two nonlinear group velocities. If $\omega_1 C_0' < 0$, the system is elliptic, and the wavetrain is unstable. For deep-water waves,

$$\omega = \sqrt{gk}(1 + \tfrac{1}{2}k^2a^2) + O(a^4);$$

hence $\omega_1 = \tfrac{1}{2}g^{1/2}k^{5/2} > 0$, $C_0' = \omega_0'' = -\tfrac{1}{4}g^{1/2}k^{-3/2} < 0$, and the wavetrain is unstable. For full details and the relation to Benjamin's analysis, see Whitham (1967a, b).

8. Formal Perturbation Theory

This final section gives a brief account of how the theory presented in Section 5 may be formally justified as the first term in a perturbation procedure. A full account is given in Whitham (1970).

The method is an interesting extension of the so-called "two-timing method," which is usually applied directly to differential equations but is here adapted to the variational principle. The method recognizes explicitly in the dependent variables that changes are occurring on two time scales: the fast oscillations of the wavetrain and the slow variations in the parameters (ω, k, a). There are two corresponding length scales. The function $u(x, t)$ is expressed in the form

$$u(x, t) = U(\theta, X, T),$$

where

$$\theta = \epsilon^{-1}\Theta(X, T), \qquad X = \epsilon x, \qquad T = \epsilon t,$$

and the small parameter ϵ measures the ratio of the fast time scale to the slow time scale; the function U no longer refers just to the periodic solution. If the wavenumber k and frequency ω are introduced, we have

$$\omega(X, T) = -\theta_t = -\Theta_T, \qquad k(X, T) = \Theta_X. \tag{91}$$

The scaling has been arranged so that

$$\frac{\partial \omega}{\partial t} = \epsilon \frac{\partial \omega}{\partial T}, \qquad \frac{\partial \omega}{\partial x} = \epsilon \frac{\partial \omega}{\partial X},$$

with similar expressions for k, and

$$\frac{\partial u}{\partial t} = -\omega \frac{\partial U}{\partial \theta} + \epsilon \frac{\partial U}{\partial T}, \qquad \frac{\partial u}{\partial x} = k \frac{\partial U}{\partial \theta} + \epsilon \frac{\partial U}{\partial X}.$$

If X, T, ω, k, U are all taken to be $O(1)$ quantities, the scaling has been arranged so that ω, k are slowly varying and so that u has a slow variation in addition to its oscillation with the phase θ.

The Euler equation for (55) is

$$\frac{\partial}{\partial t} L_1 + \frac{\partial}{\partial x} L_2 - L_3 = 0,$$

where

$$L_1 = \frac{\partial L}{\partial u_t}, \qquad L_2 = \frac{\partial L}{\partial u_x}, \qquad L_3 = \frac{\partial L}{\partial u}.$$

In the scaled variables, this becomes

$$-\omega \frac{\partial}{\partial \theta} L_1 + k \frac{\partial}{\partial \theta} L_2 - L_3 - \epsilon \frac{\partial L_1}{\partial T} + \epsilon \frac{\partial L_2}{\partial X} = 0, \tag{92}$$

where the arguments in the L_j are

$$L_j = L_j(-\omega U_\theta + \epsilon U_T, \, kU_\theta + \epsilon U_X, \, U). \tag{93}$$

Equation (92) is written as a second-order equation for the *three-variable* function $U(\theta, X, T)$. The art of the two-timing technique is to solve this equation treating θ, X, T as *independent* variables. If this can be done, then clearly $U(\epsilon^{-1}\Theta, X, T)$ is a solution of the original problem. The solution is usually obtained by means of an expansion

$$U(\theta, X, T) = \sum_{n=0}^{\infty} \epsilon^n U_n(\theta, X, T),$$

and the extra flexibility is used to suppress secular terms in θ, which would otherwise limit the uniform validity of the expansion. To lowest order in ϵ, we have

$$\frac{\partial}{\partial \theta} (- \omega L_1^{(0)} + k L_2^{(0)}) - L_3^{(0)} = 0,$$

where

$$L_j^{(0)} = L_j(- U\omega_{0\theta}, \ kU_{0\theta}, \ U_0);$$

it has the first integral

$$(- \omega L_1^{(0)} + k L_2^{(0)})U_{0\theta} - L^{(0)} = A(X, T). \tag{94}$$

Comparing with (75), we see that this is just the equation for the periodic solution but with the added dependence on (X, T). To the lowest order, therefore, we find U_0 has the same form as the periodic solution, but the parameters ω, k, A are automatically functions of (X, T) to describe the slow variations. The equations for $(\omega, \ k, \ A)$ may then be found by proceeding to the next order in ϵ. An equation for U_1 is found and the required equations for $(\omega, \ k, \ A)$ may be derived from the conditions to suppress a secular term proportional to θ in U_1.

Our interest, however, is in the variational principle. Now it is a surprising fact that (92) is the Euler equation for the variational principle

$$\delta \int \int \int_0^{2\pi} L(- \omega U_\theta + \epsilon U_T, \ kU_\theta + \epsilon U_X, \ U) \, dT \, dX \, d\theta = 0 \tag{95}$$

for the three-variable function $U(\theta, X, T)$. In (95) the function U and its variations are taken to be periodic in θ, and the variations in U vanish on the boundary of the (X, T) region. If we define

$$\bar{L} = \frac{1}{2\pi} \int_0^{2\pi} L(- \omega U_\theta + \epsilon U_T, \ kU_\theta + \epsilon U_X, \ U) \, d\theta,$$

we have already an exact form of the "averaged" variational principle. Not only is the intuitive idea sound as a first approximation, it contains the entire expansion! To lowest order,

$$\bar{L} = \mathscr{L} = \frac{1}{2\pi} \int_0^{2\pi} L(- \omega U_{0\theta}, \ kU_{0\theta}, \ U_0) \, d\theta,$$

where U_0 is the periodic solution extended to allow ω, k, A to depend on X, T as in (94). The quantity $\mathscr{L}$ is calculated as a function of (ω, k, A) alone as described in Section 5. We then have that, to lowest order, (95) is

$$\delta \iint \mathscr{L}(\omega, k, A) \, dX \, dT = 0, \tag{96}$$

where

$$\omega = -\Theta_T, \qquad k = \Theta_X.$$

The variations in Θ and A give

$$\frac{\partial}{\partial T} \mathscr{L}_\omega - \frac{\partial}{\partial X} \mathscr{L}_k = 0 \quad \text{and} \quad \mathscr{L}_A = 0,$$

respectively.

This shows the main ideas in the perturbation scheme. There are a number of points that require a more detailed explanation than is appropriate here. They are covered in Whitham (1970).

CHAPTER VI

Conservation of Wave Action

Wallace D. Hayes

Waves, in the sense used in this chapter, are dynamic oscillations of a continuous medium. They are solutions to equations of dynamics, with "dynamics" interpreted broadly. These solutions are oscillations in that (with minor reservations) they are approximately periodic in some smooth function of the independent variables termed the phase. The fact that the medium is continuous rather than discrete entails the presence of space coordinates as well as the time coordinate among the independent variables. The dynamic equations to which waves are solutions are partial differential equations.

A key problem in the theory of wave propagation is that of predicting the main properties of a wave solution at some later time in terms of an appropriate set of initial conditions. For this purpose a wave propagation theory is needed, with equations governing changes in time and space of the intensity, frequency, and wavenumber of the waves. In such a theory two basic principles appear: (1) the existence of a phase variable, whose time and distance derivatives are interpreted as the local frequency and wavenumber, and (2) a law governing the intensity, in terms of a conservation law. This chapter is concerned with the concepts underlying such a conservation law, and primarily with its analog in a discrete system.

The waves treated here are conservative in the sense that they obey equations derivable from a variational principle applied to a Lagrangian functional. We can derive some dissipative systems from a variational

The author acknowledges with thanks the support of the National Aeronautics and Space Administration under Grant NGL-31-001-119. The writing of this article was done under the tenure of an NSF Senior Postdoctoral Fellowship.

principle by pairing them with corresponding systems of negative dissipation; such systems we do not consider. In some cases a principle of conservation of energy does appear, but in general no conserved energy can be defined. For wave systems a conservation principle does appear, that of conservation of wave action. The entity action can be defined for a single wave solution only approximately, so that no absolute conservation law for a single solution appears. The conservation law is valid asymptotically with respect to a small parameter, and the entity wave action is termed an adiabatic invariant. An essential feature of the concept of action is that it involves an average over phase.

Many of the concepts that apply to waves apply also to dynamic oscillations of a discrete system. For the purpose of introducing the concept of action in the somewhat simpler context of a system with only time as the independent variable, we consider first a discrete oscillating system. This system is required to be conservative in the same sense as for the wave medium, in obeying a variational principle.

Our approach differs in detail from that of Whitham (Section 8 of Chapter V). Where Whitham averages over the phase proper, using a "two-timing" method, we average over a phase-shift parameter which determines a family of related solutions. It is important to note that this difference results in absolutely no algorithmic differences in applying Whitham's theory. In practice, there is no difference between averaging over phase and over phase shift. With our approach the basic conservation law is absolute rather than adiabatic, and the asymptotic nature of the theory lies in the identification of a family of solutions related through phase shift (Hayes, 1970a). Our approach also permits in a natural way the definition of a local action density and flux in problems in which the waves are modal or general.

1. Discrete Dynamic Systems

The time coordinate t is the independent variable. The dependent variables are denoted $\phi_i(t)$, where i is an index satisfying $1 \le i \le m$ and m is the number of distinct dependent variables. A Lagrangian function $L(\dot{\phi}_i, \phi_i, t)$ must be specified, with the dimensions of energy. The quantity $\dot{\phi}_i$ is $d\phi_i/dt$. The argument space of L encompasses $2m + 1$ variables and is termed an *extended* space. In order to distinguish derivatives taken in this space we use subscripts, while in the space of time alone we use the operator d/dt. In addition, in order to avoid summation signs, we use the standard summation convention on the index i.

According to the variational principle, the integral $\int L\,dt$ between two fixed times must be stationary with respect to variations of the dependent variables for a solution. The calculus of variations yields the m Euler equations

$$\frac{d}{dt}(L_{\dot{\phi}_i}) - L_{\phi_i} = 0. \tag{1}$$

These are the dynamic equations for the discrete system.

The Hamiltonian function H is defined by

$$H = L_{\dot{\phi}_i}\,\dot{\phi}_i - L, \tag{2}$$

using the summation convention. Using (1), we obtain

$$\frac{dH}{dt} = -L_t. \tag{3}$$

If L has no explicit dependence upon t, so that L_t is identically zero, the system is termed steady. In this case the quantity H is constant for a solution. With the Hamiltonian H identified as the total energy of the system, this result gives an energy conservation principle.

A simple special case is that of a nonlinear oscillator with spring properties varying with time. In this case there is a single dependent variable ϕ, a potential energy $G(\phi, t)$, and a restoring force (with unit mass) $g(\phi, t) = G_\phi$. The Lagrangian is

$$L = \tfrac{1}{2}\dot{\phi}^2 - G \tag{4}$$

and the equation of motion (1) is

$$\ddot{\phi} + g(\phi, t) = 0. \tag{5}$$

The Hamiltonian is

$$H = \tfrac{1}{2}\dot{\phi}^2 + G \tag{6}$$

and is not constant unless $G_t = 0$.

2. Action Conservation for a Family of Solutions

We next consider a family of solutions $\phi_i(t)$ characterized by a real parameter θ. The solutions $\phi_i(t, \theta)$ are required to be periodic in the parameter θ, and the period is chosen to be 2π. Thus we require

$$\phi_i(t, \theta + 2\pi) = \phi_i(t, \theta). \tag{7}$$

We define the *action* A associated with this family of solutions to be

$$A(t) = A[\phi_i(t, \theta)] = \overline{L_{\dot{\phi}_i} \, \phi_{i\theta}} = \frac{1}{2\pi} \oint L_{\dot{\phi}_i} \, \phi_{i\theta} \, d\theta, \tag{8}$$

where the bar indicates average over a period of θ (with the summation convention). The quantity A is a functional of the family of solutions chosen and for that family is a function of t.

Direct calculation of dA/dt, using (1), yields the result

$$\frac{dA}{dt} = \overline{L_\theta} = 0. \tag{9}$$

Thus $A(t)$ is a constant, an absolute invariant associated with the given family of solutions. It has the dimensions of energy times time (and divided by the dimensions of θ if any have been assigned to it).

This result, as it stands, is quite weak. It gives us no information about any single solution. Only to the extent that we are able to identify a family of solutions with a single solution of interest will we be able to apply the result to a single solution.

3. Steady Oscillating System

We consider next an oscillating solution of a steady system (one with $L_t \equiv 0$). The solution is periodic in a phase variable ϑ which depends linearly on the time, with period 2π. Thus, the solution is of the form

$$\phi_i(t) = \Phi_i(\vartheta) = \Phi_i(\vartheta + 2\pi). \tag{10}$$

The phase is taken of the form

$$\vartheta = -\omega t - \theta, \tag{11}$$

with ω the angular frequency of the oscillation and θ a constant. Since the system is steady, the solution remains a solution if an arbitrary constant is added to the time. Since the solution depends functionally only on ϑ, we see from (11) that the solution remains a solution under an arbitrary change of θ. Thus θ, here interpreted as a phase shift, has the properties specified in Section 2. The single periodic solution has generated a family of solutions through phase shift.

The action A calculated for the solution family can be associated with the single solution and can be considered a measure of the intensity of the oscillation. The invariant (in this case) Hamiltonian may be calculated; it is

$$\bar{H} = H(A) = \omega A - \mathscr{L}, \tag{12}$$

where $\mathscr{L} = \bar{L}$ is the phase average of the Lagrangian for the solution. Another result may be obtained. From (1) we may show that $d(L_{\dot{\phi}_i} \phi_i)/dt = L_{\dot{\phi}_i} \dot{\phi}_i + L_{\phi_i} \dot{\phi}_i$. Integrating this over a single period yields

$$\overline{L_{\dot{\phi}_i} \phi_i + L_{\phi_i} \dot{\phi}_i} = \frac{1}{2\pi} \oint (L_{\dot{\phi}_i} \dot{\phi}_i + L_{\phi_i} \dot{\phi}_i)\, d\theta = 0. \tag{13}$$

In the special case for which L is a quadratic form in the variables $\dot{\phi}_i$ and ϕ_i, the basic equations (1) are linear and homogeneous in ϕ_i. For such a linear system $L_{\dot{\phi}_i} \dot{\phi}_i + L_{\phi_i} \phi_i = 2L$, (13) may be read that $\bar{L} = \mathscr{L} = 0$, and (12) that $H = \omega A$. With L interpreted as kinetic energy less potential energy, this is the classical equipartition result for linear oscillators, which states that the averages of kinetic and potential energies are equal. For a general nonlinear oscillating system (13) is the analogous result.

The discussion above has been for a single solution to (1) of the form (10), with specific values of ω and of A. Such a single solution is not isolated, in general, and there are neighboring periodic solutions with slightly different values of ω and of A. Such neighboring solutions are said to be of the same oscillation mode. For a particular mode the quantities ω and A are interdependent, in general, and we may write

$$\omega = W(A) \tag{14}$$

to indicate the functional dependence of the frequency on the intensity of the oscillation. For the corresponding family of solutions we may write in place of (10)

$$\phi_i = \Phi_i(\vartheta, A). \tag{15}$$

In Whitham's approach to problems of this type (see Chapter V) the average Lagrangian $\mathscr{L}$ is calculated from (11) and (15) as a function $\mathscr{L}(\omega, A)$ without requiring that ω satisfy (14). The variational principle is then applied to the average Lagrangian with respect to variations in intensity A and phase ϑ. With this approach the variation with respect to A gives

$$\mathscr{L}_A(\omega, A) = 0 \qquad (16)$$

as the equivalent of (14). Variation with respect to ϑ gives

$$\frac{d\mathscr{L}_\omega}{dt} = 0$$

as the equivalent of (9). The derivative $\mathscr{L}_\omega$ with ω assigned the correct value by (14) or (16) may be identified with A for a solution. With $L(\omega, A)$ given by (12), the result $\mathscr{L}_\omega = A$ is immediate. From (12) and (16) appears also the specific result that (14) is expressible as

$$\omega = W(A) = \frac{dH}{dA}. \qquad (14')$$

Returning to the example of the nonlinear oscillator discussed in Section 1 above, the steady oscillating solution requires that $G_t = 0$. The solution is given by

$$t = \int \frac{d\phi}{[2(H - G)]^{1/2}}.$$

The frequency is given by $\omega = 2\pi/T$, where T is the period

$$T = \oint \frac{d\phi}{[2(H - G)]^{1/2}}.$$

The action may be calculated from (8):

$$A = \frac{1}{2\pi} \oint [2(H - G)]^{1/2} \, d\phi.$$

The result (14') may be readily checked. The action A equals the area within the closed curve representing the solution on a plane with cartesian variables ϕ and $L_{\dot\phi} = \dot\phi = [2(H - G)]^{1/2}$, times $1/2\pi$.

4. Action Conservation for a Slowly Varying System

The constancy of A for an oscillating solution for a strictly steady system is, of course, trivial. A very slowly varying system can be formulated by replacing the explicit dependence upon t in the Lagrangian L by a dependence upon a slow time T given by

$$T = \epsilon t, \tag{17}$$

with ϵ a small parameter. With ϵ very small the system behaves approximately as a steady system with T considered to be a constant independent parameter.

An approximate quasiperiodic solution of the form

$$\phi_i(t) = \Phi_i(\vartheta, A, T) \tag{18}$$

in place of (15) may be sought, with $A = A(T)$, $\omega = \Omega(A, T)$, and

$$\vartheta = -\int W(A, T) \, dt - \theta \tag{19}$$

in place of (11). The phase-shift parameter θ in (19) then yields the family of solutions required to define the action A. The basic result (9) then reads $dA/dT = 0$, or $A(T) = const$. This then provides a nontrivial action conservation law for this oscillatory solution to our slowly varying system. In the example of the nonlinear oscillator, now with spring properties varying very slowly with time, this result says that the intensity of the oscillation (as measured by H or by ϕ_{max}, for example) adjusts itself continuously so that the quantity A remains constant.

This argument is disarmingly simple and direct, but subject to certain pitfalls and caveats. The argument obviously fails if the approximate solution fails to exist, bifurcates, or becomes singular in some other way at some value of T. We simply must assume this is not the case, that periodic solutions of the type (18) do exist for all values of T with T frozen as a parameter and are identifiable as corresponding to the same mode, and we assume we can show that approximate solutions of the type (18) exist with $T = \epsilon t$ slowly varying. The more subtle questions concern the nature of the approximations involved.

To explore this point let us consider a system that is steady for $T < 0$, with a one-parameter family of periodic solutions of the type considered in Section 3 strictly established. At $T = 0$ the system becomes slowly varying, and at $T = T_0$ the system again becomes steady. The shifts in type involve nonanalytic behavior of the functions describing the system at $T = 0$ and $T = T_0$, and it is assumed that these nonanalyticities are gentle. For $T > T_0$ the quantity A, defined according to Section 2, equals its original value because A is an absolute invariant.

The approximate nature of the result lies in the fact that for $T > T_0$ the family of solutions parameterized by the original phase-shift parameter is no longer strictly of the form (15) or (18). In the example of the non-linear oscillator (4), (5), each solution in the family may be periodic and thus may be identified with an action $A(\theta)$ and a frequency (14) according to Section 3. In general, however, $A(\theta)$ and $\omega(\theta)$ are not strictly constant. The absolute invariant A defined for the family is in some appropriate sense an average of $A(\theta)$ over θ.

In systems with more degrees of freedom another difficulty appears. The solution, strictly periodic for $T < 0$, is, in general, not periodic for $T < T_0$, or is periodic with a period much larger than that corresponding to the function $W(A, T)$. In a linear system this may be interpreted in terms of an admixture of another normal mode, or mode mixing. In the general nonlinear case we use the same term for this phenomenon. The solution may be identified as a periodic solution plus a small perturbation.

The errors of these two types are small if ϵ is small. In the limit $\epsilon \to 0$, in which the slow variations become infinitely slow, the errors disappear. Thus the results—the preservation of the periodicity of a solution and the invariance of an action A defined for a particular family of solutions—are termed asymptotic, and such an A is termed an adiabatic invariant. The precise manner in which the errors depend functionally upon ϵ depends upon the nature of the nonanalyticities at $T = 0$, T_0. The errors delimit our ability to identify the solution as a periodic oscillation.

An alternative procedure not involving the concept of action is to use an average Hamiltonian $\bar{H}$ averaged over phase in place of A as a measure of oscillation intensity. The derivative L_t becomes ϵL_T. Equation (3) may be averaged over phase to yield

$$\frac{d\bar{H}}{dT} = -\bar{L}_T. \tag{20}$$

With the righthand side expressed as a function of $\bar{H}$ and T, equation (20) is a differential equation that may be solved for $\bar{H}(T)$. This procedure, for several reasons, is less desirable than that of using action. The quantity $\bar{H}$ is less fundamental than is A. All the asymptotic errors that appear with A appear also with $\bar{H}$. The procedure with $\bar{H}$ does not rest upon a basic absolute conservation law. Finally, the differential equation (9) for A is immediately integrated to the result $A = const$.

In general, whichever method is used, the essence of the application to slowly varying systems lies in the ability to average over phase. A solution must, except for the phase itself, propagate in a manner independent of phase. The phase must be ignorable. The errors in the results are a direct measure of the extent to which specific phase relationships affect the solution.

5. Continuous Media

The dynamic oscillations that are waves occur in a continuous medium. The medium is assumed to be conservative in the same sense as is the discrete system considered above, in that its dynamic equations are derivable from a variational principle. The independent variables are now $\mathbf{x}$, t, with the components of the vector $\mathbf{x}$ distance coordinates. The dependent variables are designated as before but have a functional dependence $\phi_i(\mathbf{x}, t)$. The gradient $\nabla\phi_i$ and $\dot{\phi}_i$ are the first derivatives of ϕ_i in its physical space of dependence.

A Lagrangian density $L(\dot{\phi}_i, \nabla\phi_i, \phi_i, \mathbf{x}, t)$ must again be specified. Again, derivatives in physical space are denoted by the operators ∇ and $\partial/\partial t$ to distinguish them from derivatives in the extended argument space of L, designated by subscripts. The variational principle requires that the integral $\iint L\, dt\, d\mathbf{x}$ be stationary for a solution ($\int d\mathbf{x}$ is a short notation for $\iiint dx_1\, dx_2\, dx_3$). The calculus of variations yields the m Euler equations analogous to (1):

$$\frac{\partial}{\partial t}(L_{\dot{\phi}_i}) + \nabla \cdot (L_{\nabla\phi_i}) - L_{\phi_i} = 0. \tag{21}$$

These partial differential equations are the dynamic equations for the continuous medium. Note that $L_{\nabla\phi_i}$ is a vector quantity, with x_j component equal to $L_{\partial\phi_i/\partial x_j}$.

A Hamiltonian density may be defined using (2). It obeys the equation analogous to (3):

$$\frac{\partial H}{\partial t} + \nabla \cdot (L_{\nabla\phi_i}\dot{\phi}_i) = -L_t, \tag{22}$$

using, of course, the summation convention. If L has no explicit dependence upon t, with the system steady this takes the form of what is termed in continuum theories a conservation equation. Such an equation states that the time derivative of a density plus the divergence of a flux vector equals zero. The vector $L_{\nabla\phi_i}\dot{\phi}_i$ is to be interpreted as a Hamiltonian flux vector. The reason for the designation "conservation equation" is discussed below in connection with action density.

The analysis of Section 2 extends directly to continuous media. A family of solutions $\phi_i(\mathbf{x}, t, \theta)$ parameterized by the parameter θ and periodic of period 2π in θ is considered. The *action density* $A(\mathbf{x}, t)$ is defined by (8) (with functional dependence upon $\mathbf{x}$ also, of course). The *action flux* $\mathbf{B}(\mathbf{x}, t)$ is defined by

$$\mathbf{B}(\mathbf{x}, t) = \mathbf{B}[\phi_i(\mathbf{x}, t, \theta)] = \overline{L_{\nabla\phi_i}\phi_{i\theta}} = \frac{1}{2\pi} \oint L_{\nabla\phi_i}\phi_{i\theta}\, d\theta. \tag{23}$$

The analog of (9) is the conservation equation

$$\frac{\partial A}{\partial t} + \nabla \cdot \mathbf{B} = 0. \tag{24}$$

The total action at any time is the integral $\int A\, d\mathbf{x}$ of the action density over the domain. It is assumed that at a given time this integral converges and the total action is well determined. It is further assumed that any boundaries to the domain are fixed (not moving) and that on these boundaries, for each value of the index i, either $\phi_i = 0$ or $\mathbf{n} \cdot L_{\nabla\phi_i} = 0$; here $\mathbf{n}$ is a unit normal vector. If the domain is infinite, it is assumed that the solution approaches zero sufficiently rapidly as $|\mathbf{x}| \to \infty$. Direct calculation of the time derivative of this integral using (24) and the divergence theorem, with a limiting process if the domain is infinite, leads to

$$\frac{d}{dt} \int A\, d\mathbf{x} = 0. \tag{25}$$

The conclusion is then drawn that total action is conserved. This conclusion depends upon the "divergence form" of (24) and explains the appellation "conservation equation" for an equation of this form.

6. Waves

Waves, considered as oscillations of a continuous medium, are solutions to the dynamic equations that are periodic in a phase variable ϑ. This phase is a function $\vartheta(\mathbf{x}_{\|}, t)$ in a *propagation space*, which is not, in general, the same as the physical space $(\mathbf{x}, t)$. The physical space is considered to be representable as the product of the propagation space and a *cross* or *lateral space* $(\mathbf{x}_{\perp})$.

If the medium is steady and uniform, by which is meant that L has no explicit dependence upon t or $\mathbf{x}_{\|}$, we can expect strictly periodic solutions (of period 2π) in ϑ with ϑ having the form

$$\vartheta = \mathbf{k} \cdot \mathbf{x}_{\|} - \omega t - \theta. \tag{26}$$

Such solutions form the analogy to the strictly periodic solutions of form (10), (11) for a discrete system. They correspond to waves that are strictly plane in the propagation space. The vector $\mathbf{k}$ is the wavenumber.

More generally, and particularly if the medium is slowly varying (with an explicit dependence upon $T = \epsilon t$ and $\mathbf{X} = \epsilon \mathbf{x}_{\|}$ in the Lagrangian density), the phase ϑ must be a more general function of $\mathbf{x}_{\|}$ and t. We define frequency and wavenumber then by the relations

$$\omega(\mathbf{X}, T) = -\frac{\partial \vartheta}{\partial t} \tag{27a}$$

and

$$\mathbf{k}(\mathbf{X}, T) = \nabla_{\|} \vartheta. \tag{27b}$$

The frequency and wavenumber thus defined must obey the consistency relation

$$\frac{\partial \mathbf{k}}{\partial t} + \nabla_{\|} \omega = 0. \tag{28}$$

This equation is identified by Whitham as guaranteeing conservation of wave crests. It is in the form of a vector conservation equation, with density $\mathbf{k}$ and flux $\omega \mathbf{I}$, where $\mathbf{I}$ is the idemtensor.

If the cross space is a point (and hence trivial), so that the propagation space is the same as the physical space, wave solutions are termed of *local* type. If the cross space is not trivial, two cases must be distinguished: (1) Homogeneous boundary conditions on a fixed boundary in the cross space, or appropriate evanescence conditions if the cross space is not bounded, are imposed; here the waves are termed of *modal* type. (2) Such conditions are either partially applied or not at all; here the waves are termed of *general* type. The principal distinction between the two cases lies in the property of the modal case that an eigenvalue-eigenfunction problem appears; solutions of the type desired do not exist for arbitrarily specified ω, $\mathbf{k}$, and amplitude. Waves of either local or modal type are termed *plane* waves or progressive waves, in that they are approximately plane in the propagation space.

For any type of wave, the approximate periodicity in a phase ϑ permits the establishment of a family of approximate solutions that differ only by a phase shift. An action density A and flux $\mathbf{B}$ in the physical space may be then defined, subject to errors of identification completely analogous to those for discrete systems, and obeying the basic conservation law (24).

The solutions analogous to the strictly periodic solutions of Section 3 are those in a medium which is steady and uniform and of local or modal type that are strictly periodic functions $\phi_i(\vartheta, \mathbf{x}_\perp)$ of the phase defined in (26). For such solutions analogs of the results of Section 3 appear. The analog of the generalized equipartition law (13) is

$$\int_\perp \overline{(L_{\phi_i}\dot{\phi}_i + L_{\nabla\phi_i} \cdot \nabla\phi_i + L_{\phi_i}\phi_i)} \, d\mathbf{x}_\perp = 0. \tag{29}$$

In the linear case, with L quadratic in ϕ_i and its derivatives, this says that $\mathcal{L} = \int_\perp \overline{L} \, d\mathbf{x}_\perp = 0$. The appropriate homogeneous boundary conditions or evanescence conditions are needed to prove (29) in the modal case.

Another immediate result is that

$$\nabla_\perp \cdot \mathbf{B}_\perp = 0, \tag{30}$$

obtained from (24) by observing that A and $\mathbf{B}$ are independent of t and $\mathbf{x}_\|$.

We define the action density $\mathscr{A}$ and flux $\mathscr{B}$ in the propagation space by

$$\mathscr{A} = \int_{\perp} A \, d\mathbf{x}_{\perp} \tag{31a}$$

and

$$\mathscr{B} = \int \mathbf{B}_{\parallel} \, d\mathbf{x}_{\perp}. \tag{31b}$$

The analog of (14) is the nonlinear dispersion relation

$$\omega = W(\mathscr{A}, \mathbf{k}). \tag{32}$$

In the analog of (15) the dependence upon A is replaced by one on $\mathscr{A}$ and $\mathbf{k}$. In Whitham's approach (Chapter V) the equivalent of (32) is $\mathscr{L}_{\mathscr{A}} (\omega, \mathscr{A}, \mathbf{k}) = 0$, and $\mathscr{L}_{\omega}$ is identified as $\mathscr{A}$.

In a slowly varying system the solutions are again considered very close to the strictly periodic solutions of a uniform medium. The conservation equation (24) is integrated over the cross space to give the basic conservation law

$$\frac{\partial \mathscr{A}}{\partial t} + \nabla_{\parallel} \cdot \mathscr{B} = 0 \tag{33}$$

in the propagation space.

If $\mathscr{A}$ and $\mathscr{B}$ are defined by (31) for a one-parameter family of exact solutions, (33) is an exact conservation law. As in the case of the discrete system, the errors that may appear in the application of (33) are primarily those in the ability to identify a solution as a periodic solution, or rather in terms of periodic solutions. Accordingly, $\mathscr{A}$ and $\mathscr{B}$ are the density and flux of an adiabatically invariant entity defined for a wave system.

As compared with the discrete system, a continuous medium can exhibit a far greater diversity of pathological behavior. In many cases, however, a wavelike solution in an infinite propagation space for large time tends to separate into wave packets which are more or less individually identifiable. In such cases, the initially identifiable total action at large time equals the sum of the actions to be associated with the distinct wave packets.

Our account of wave action has ignored the possible effect of what Whitham (Chapter V) terms "potential variables." A potential variable is a dependent variable ϕ_i that appears in the Lagrangian L only in terms of its derivatives $\dot{\phi}_i$ and $\nabla\phi_i$. Such variables need a slightly separate treatment. They can be important in fluid-mechanical waves, for example, where there is an essential interaction between a wave system and a mean flow system.

7. Comparison of Action and Energy Conservation

An obvious question is why we should introduce the relatively unfamiliar concept of action, definable only for a family of solutions, in place of the much more familiar concept of energy. Energy is conserved in physical systems and can be defined for a single solution. Energy averaged over phase can be used in place of action in many cases, particularly where the medium is steady and $L_t = 0$.

The limitations in the application of the concept of energy conservation for waves appears most clearly in fluid or plasma dynamics. In these cases the undisturbed medium, with no waves present, is itself a solution of the equations of motion which may not be trivial. The energy for the entire undisturbed system may be infinite. In any case, waves are to be considered as a perturbation solution, for which a perturbation Lagrangian may be needed to yield the appropriate perturbation equations of motion satisfied by the waves. The energy or action used in the study of waves should characterize the waves and thus may need to be defined as perturbation entities. Perturbation energy is conserved only when the undisturbed solution is steady.

In order to use energy conservation for waves, we must average the energy density over phase. There is little essential distinction between averaging over phase and over phase shift (θ), and the problems and approximations of the two procedures are about the same. There is no advantage to be gained from the fact that energy can be defined for a single solution instead of for a family of related solutions.

Under a galilean transformation with velocity U, A and $\mathscr{A}$ are unaltered, while B and $\mathscr{B}$ are altered by additive terms UA and $U\mathscr{A}$. The energy defined as a Hamiltonian does not follow such simple rules. This lack of proper behavior for the Hamiltonian, however, could have been foreseen. A galilean transformation applied to a slowly varying steady medium, in which energy is conserved, leads generally to an unsteady medium, for which energy is not conserved.

The equations of motion for a physical system are unchanged by the addition to L of a term equal to a derivative with respect to t or a cartesian component of $\mathbf{x}$ of an arbitrary function of the form $M(\phi_i, \mathbf{x}, t)$. Such a transformation, termed a gauge transformation, must be permitted, since it does not change the problem in any real way. Action densities and fluxes are invariant under such a transformation; Hamiltonian energy density and flux are not.

As a general philosophical rule in physics and applied mathematics it is helpful to investigate behavior under various types of transformations and to hold as more fundamental those entities that are appropriately invariant. For waves, this rule points to action as a more fundamental entity than energy.

8. Wave Propagation Theory

The principal problem in a study of the propagation of progressive waves is that of predicting the course of the wave intensity (characterized by A or $\mathscr{A}$) and the frequency and wavenumber (characterized by $\mathbf{k}$). For simplicity now we assume the waves are local, so that $\mathbf{x} = \mathbf{x}_{\parallel}$ and $A = \mathscr{A}$. The basic governing equations are (28) and (24) or (33), subject to the functional behavior (32) of $\omega = W(A, \mathbf{k}, \mathbf{x}, t)$, and

$$\mathscr{B} = \mathscr{B}(A, \mathbf{k}, \mathbf{x}, t),$$

characteristic of a particular mode.

The basic equations (28) and (24) or (33) are equations in the propagation space $(\mathbf{x}, t)$. The argument space of the functions W and $\mathbf{B}$ is an *augmented* space $(A, \mathbf{k}, \mathbf{x}, t)$. With the functional behavior in the augmented space introduced into the basic equations they become

$$\frac{\partial \mathbf{k}}{\partial t} + W_{\mathbf{k}} \cdot \nabla \mathbf{k} + W_A \nabla A = - W_{\mathbf{x}} \tag{34}$$

and

$$\frac{\partial A}{\partial t} + \mathbf{B}_A \cdot \nabla A + \mathbf{B}_{\mathbf{k}} : \nabla \mathbf{k} = - \operatorname{tr}(\mathbf{B}_{\mathbf{x}}). \tag{35}$$

The terms on the righthand sides of these equations represent the slow variations in the medium and are zero in a uniform medium.

It may be shown that $W_{\mathbf{k}} = \mathbf{B}_A$, and this vector we term the *basic grou velocity*. It may also be shown that the tensor $\mathbf{B}_k$ is symmetric. In the case of linear wave propagation $W = W(\mathbf{k}, \mathbf{x}, t)$ and the term in W_A disappears from (34). The basic group velocity is then the ordinary *group velocity*. Equation (34) is decoupled from (35) and can be directly solved by the method of characteristics. The characteristics in this case are the rays $d\mathbf{x}/dt = W_k$ of standard theories of kinematic wave theory (including, for example, geometric optics and geometric acoustics). (See Hayes 1970b and references cited therein.) After the rays have been found and $\mathbf{k}(\mathbf{x}, t)$ is known, equation (35) may be solved to find the wave intensities.

In nonlinear wave propagation, (34) and (35) are coupled, and (34) cannot be solved without solving (35) simultaneously. Such problems are beyond the scope of this chapter but are partly treated in Chapter V.

Bibliographical Note and References

The principal development of the theory with its application to a number of various wave problems is due to Whitham (1965a, 1965b, 1967a, 1967b). He has also presented (1970) a careful mathematical justification of his methods and results. The approach defining action density and flux as local averages over a phase-shift parameter will be found in Hayes (1970a).

The first general formulation in print of the general wave equations for linear wave motion appeared by Whitham (1961). A review of the concept of group velocity and reference to much early work will be found in Lighthill (1965a).

Nonlinear wave propagation theory is explored in the papers of Whitham cited and also in Lighthill (1965b). See also Hayes (1973).

CHAPTER VII

Wave Interactions

Owen M. Phillips

Wave motions in nature often do not appear as a single train of periodic or almost periodic waves but assume a much less regular form. A familiar example is ocean waves, whose irregularity results both from their generation over a considerable area and from the intrinsic instabilities in the wave motion itself. In motions of this kind, an individual Fourier component can, to an extent, be regarded as propagating across the water surface, interacting with other wave components (and with itself) as it does so. The nature of the interactions among individual wave components (the primary subject of this chapter) has important ramifications in a variety of physical contexts. The basic ideas are quite simple and can be perceived most readily by considering the nature of dispersive waves in general. Details will be discussed for water waves; but it should be borne in mind that the methods are general and find application in other problems.

Suppose p represents some property of the motion—for example, the displacement of a fluid element from its equilibrium position, or the fluid pressure; then certain types of wave motion, such as infinitesimal sound waves, are governed by the classical linear wave equation

$$p_{tt} - c^2 \nabla^2 p = 0,$$

where c is the speed of sound. This equation has elementary solutions of the type

$$p = a \exp i\mathbf{k} \cdot (\mathbf{x} \pm ct),$$

describing the propagation of a wavetrain with a definite speed c independent of the wavelength. These waves are, of course, nondispersive. However, many of the waves encountered in fluid systems do not obey the classical wave equation, and their phase velocity may not be independent of the wavenumber. More generally, for an infinitesimal disturbance, the governing equation is generally of the type

$$\mathscr{L}(p) = 0, \tag{1}$$

where $\mathscr{L}$ is a linear operator involving derivatives with respect to position and time. The form of this operator depends upon the particular type of wave motion considered. For example, infinitesimal surface waves, for which the surface boundary condition provides a wave equation

$$\phi_{tt} + g\phi_z = 0 \qquad \text{at } z = 0$$

for the velocity potential, are discussed in Chapter IV, Section 4. Another example is internal gravity waves in a stratified fluid with constant Brunt-Väisälä frequency N (see Chapter X, Section 1), for which we have

$$\nabla^2 w_{tt} + N^2 \nabla_h^2 w = 0, \tag{2}$$

where w is the vertical velocity of fluid elements and ∇_h^2 the horizontal Laplacian operator. Again, these linear equations admit solutions of the form

$$\phi, \ w = a \exp i \ (\mathbf{k} \cdot \mathbf{x} \pm \omega t),$$

provided that w and k satisfy a dispersion relation that is characteristic of the type of wave motion considered.

For gravity waves on deep water the dispersion relation is

$$\omega^2 = gk, \tag{3}$$

while for the internal gravity waves specified by equation (2)

$$\omega = N \cos \theta, \tag{4}$$

where θ is the angle between the wavenumber and the horizontal direction.

The very fact that these infinitesimal wave solutions are useful at all suggests that in many instances the interaction and other nonlinear effects are weak; the infinitesimal amplitude solution represents in many cases a useful first approximation to the motion. In the historical development of fluid mechanics, the study of surface waves on water provided perhaps the first instance where theoretical solutions compared favorably with observational experience. However, all such equations of the type (1) in fluid mechanics represent an approximation to a nonlinear equation in which the nonlinear terms are, in some sense, small. Our expectation that the interaction effects that do arise from these nonlinear terms will, in most circumstances, be weak, suggests that we might profitably examine their general nature by considering an equation of the form

$$\mathscr{L}(p) = \epsilon \mathscr{N}(p), \tag{5}$$

where $\mathscr{N}$ is some nonlinear operator, whose precise nature again depends upon the particular wave system considered, and ϵ is a small parameter characterizing the amplitude of the motion. Solutions to an equation of this type can be heuristically deduced by successive approximation—in essence, by expanding in powers of ϵ. The first approximation is the linear solution; this is substituted into the righthand side, and the resulting equation is solved to obtain second approximations, and so on. Refinements of this scheme will emerge in due course.

1. The Resonance Conditions

By way of background, let us consider a very elementary property of oscillating systems. A linear oscillator subjected to an infinitesimal forcing function is described by the equation

$$x_{tt} + k^2 x = \epsilon e^{i\Omega t},$$

where k is the natural frequency of the system and Ω the frequency of the forcing function. The response of the system from an initial state of rest is small (of order ϵ) unless k^2 is very nearly equal to Ω^2—that is, unless there is resonance between the frequency of the forcing function and the natural frequency of the system. If the two are precisely equal, the amplitude of the oscillation grows linearly with time and becomes arbitrarily large. Indeed, the only way that the response can become large is through the phenomenon of *resonance*.

A very similar effect is involved in the interaction among wavetrains. Let us consider two interacting wavetrains, given to a first approximation by solutions to the linear equation (3):

$$a_1 \exp i(\mathbf{k}_1 \cdot \mathbf{x} - \omega_1 t),$$
$$a_2 \exp i(\mathbf{k}_2 \cdot \mathbf{x} - \omega_2 t), \tag{6}$$

where the individual wavenumbers and frequencies are related by the dispersion relation $\omega_i = W_i(\mathbf{k}_i)$, $i = 1, 2$. The solution to the complete equation (5) would be expected heuristically to be of the same general form, though the amplitude a may possibly vary with time. We might seek solutions to (5) by substituting expressions of the type (5) into the small nonlinear terms on the right. If the lowest-order nonlinearity is quadratic, this substitution leads to expressions of the type

$$\exp i[(\mathbf{k}_1 \pm \mathbf{k}_2) \cdot \mathbf{x} - (\omega_1 \pm \omega_2)t].$$

Now, in (5) these terms act as a small amplitude forcing function to the essentially linear system and provide an excitation at the wavenumbers $(k_1 \pm k_2)$ and frequencies $(\omega_1 \pm \omega_2)$. The response of the system to this forcing function can be expected to be small (that is, of order ϵ) unless resonance occurs—that is, unless the wavenumber and frequency at which the forcing is applied correspond to a (wavenumber, frequency) pair of a natural wave mode. If they do correspond, we have resonance and growth of this new component with continuous energy transfer from the two original wave components to this third one. If the initial amplitude of the third component is zero, then under resonant conditions it grows linearly until the energy drain from the other components begins to reduce their amplitudes and consequently the amplitude of the forcing function. Although the basic idea is a simple extension of the well-known behavior of a linear oscillator, some mathematical niceties occur, particularly when we consider the almost, but not quite, resonant case. These have been discussed by Bretherton (1964).

It appears, then, that in quadratic interactions the conditions under which three components can exchange energy among themselves can be written

$$\omega_1 \pm \omega_2 \pm \omega_3 = 0,$$
$$\mathbf{k}_1 \pm \mathbf{k}_2 \pm \mathbf{k}_3 = 0, \tag{7}$$

in which, for each wavenumber frequency pair, the linear dispersion relation

$$\omega_i = W_i(\mathbf{k}_i), \qquad i = 1, 2, 3$$

holds. If the three components do not satisfy these conditions, the product of their interaction is small, of order ϵ.

The existence of energy transfer of this kind clearly depends upon the existence of solutions to these resonance conditions. This, in turn, depends upon the form of the dispersion relation, which is a characteristic of each type of wave motion. For example, with capillary-gravity waves on the surface of deep water, the existence of triads of wavenumbers that satisfies the resonant conditions can be shown most simply by a geometrical construction (McGoldrick, 1965).

First, let us consider only unidirectional waves and plot the frequency as the ordinate, the wavenumber as the abscissa. For gravity-capillary waves, the dispersion relation is

$$\omega = \left(gk + \frac{Tk^3}{\rho} \right)^{1/2}, \tag{8}$$

where T is the surface tension and ρ the water density. This is shown as the solid curve in Figure VII.1. For small values of k, in the gravity-wave range, the curve is concave downwards; in the capillary range at larger values of k it is convex downwards. If we chose any point (ω_1, k_1) on this curve and draw the vector (such as 1) connecting it to the origin, then the vertical and horizontal components give a wave-number, frequency pair for free waves. Now, with (ω_1, k_1) as the origin, construct a new coordinate system with axes ω' and k' and plot the same dispersion curve in these new coordinates. This is shown in Figure VII.1 as the broken line. In a similar manner, any vector drawn between a point on the broken curve and the second origin also specifies a wavenumber, frequency pair for free waves. If the two curves intersect [as they do in this case at (ω_3, k_3), which is also $\omega' - \omega_2$, $k' - k_2$] then the three vectors joining the intersection point with the two origins, and the two origins themselves, specify a triad of wavenumbers and frequencies satisfying (7) and, for each pair separately, (8). In the example shown,

$$\omega_1 + \omega_2 = \omega_3,$$

$$k_1 + k_2 = k_3.$$

These three components are capable of resonant interaction among themselves. Note that these interactions are selective, since three wavenumbers chosen arbitrarily do not, in general, satisfy these relations. They are also weak because of the smallness of the nonlinear term; we see later that it takes wave periods of the order of $(\epsilon)^{-1}$ for significant transfer to take place.

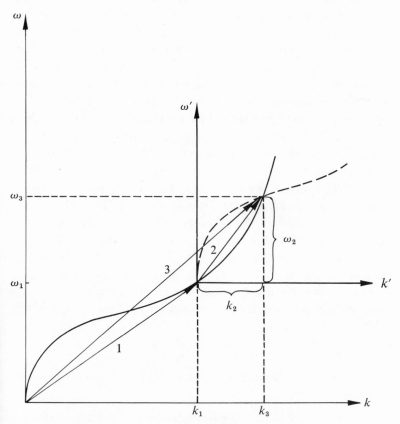

VII.1. A diagram in the wavenumber-frequency plane illustrating the existence of resonant triads in gravity-capillary waves.

This construction does no more than assure us that at least some collinear capillary-gravity waves are capable of such resonant interactions. If the wavetrains are no longer collinear, the wavenumbers **k** are horizontal vectors. The geometrical construction must then be made in three dimensions, and the dispersion relation is no longer a curve but a cusped bowl, axially symmetric about the frequency axis. The intersections of two such bowls define the wavenumber triads capable of resonant interaction (McGoldrick, 1965). These are illustrated in Figure VII.2.

In this particular type of wave motion, then, triads of components do exist that can interact resonantly, but this is not always so. For example, with pure gravity waves on the surface of deep water, the dispersion relation is (3) and the corresponding dispersion curves are everywhere concave downwards. A similar construction gives no intersections, so that there are no nontrivial solutions to the resonance conditions for quadratic interactions. In this case, one must continue the analysis to a higher order, to discover that the first resonant effects occur with cubic interactions involving four separate wavenumbers; this is a complication that will be discussed later. For many types of wave motion, however, the simpler case of quadratic resonant interactions among a triad of wavenumbers may occur. This is the case for internal gravity waves in a continuously stratified fluid with constant Brunt-Väisälä frequency N and dispersion relation (4), and in (the closely analogous) inertial waves in a uniformly rotating fluid.

Resonance can also occur among the components of two coexisting types of wave motions. For example, Ball (1964) and Thorpe (1966) have shown that the interaction of two gravity waves moving over a water surface can, in principle, generate internal waves on a sharp, submerged thermocline.

2. The Interaction Equations

Let us now examine the dynamics of a set of three wavetrains undergoing resonant interactions. Let

$$p = \sum_i \{a_i(\epsilon t) \exp [i(\mathbf{k}_i \cdot \mathbf{x} - \omega_i t)] + a_i^*(\epsilon t) \exp [-i(\mathbf{k}_i \cdot \mathbf{x} - \omega_i t)]\},$$

where $i = 1, 2, 3$. Note that p is necessarily real. The wave amplitudes a

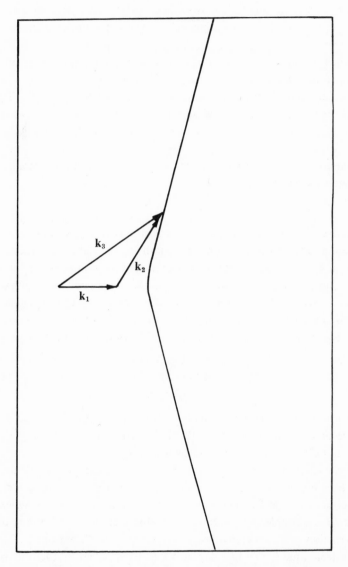

VII.2. A diagram in the wavenumber plane showing a typical resonant triad for capillary-gravity waves, following McGoldrick (1965). The wavenumber k_1 is capable of interaction with any wavenumber k_2 terminating on the line shown, with k_3 completing the triangle.

may be slowly varying functions of time, and it is this variation that we wish to examine by substitution of an expected solution of this kind into (5).

When this substitution is made, the linear operator $\mathscr{L}$ acting on the exponential terms vanishes precisely, since the wavenumbers and frequencies satisfy the dispersion relation associated with (1). Time derivatives of the linear operator will result in first derivatives of a of order ϵ, second derivatives of order ϵ^2, and so on. On the righthand side, the quadratic terms will result in sums and differences of the wavenumbers and frequencies concerned, and if, for example, the resonance is such that

$$X_1 + X_2 + X_3 = 0, \qquad X_i = \mathbf{k}_i \cdot \mathbf{x} - \omega_i t, \tag{9}$$

then the term on the lefthand side proportional to $\exp iX_1$ has the same spatial and temporal variation as the term on the right proportional to $\exp -i(X_2 + X_3)$. These terms can be extracted by multiplying the whole equation by the complex conjugate of these factors and averaging over a few wavelengths. In this way, one separates dynamical equations specifying the time rate of change of the three amplitudes concerned. The equations that result are, to the lowest order in ϵ, of the form

$$
\begin{aligned}
i\dot{a}_1 &= C_1 a_2{}^* a_3{}^*, \\
i\dot{a}_2 &= C_2 a_3{}^* a_1{}^*, \\
i\dot{a}_3 &= C_3 a_1{}^* a_2{}^*.
\end{aligned}
\tag{10}
$$

These are the interaction equations for a resonant trio, derived much more precisely by Bretherton (1964). The constants C_i are, in general, functions of the geometrical configuration of the wavenumbers and of the particular wave type involved. In specific problems, the calculation of these interaction coefficients by the elementary substitution method described here is frequently a matter of great algebraic complexity. Very much simpler but more sophisticated methods of obtaining these interaction coefficients involve the use of Whitham's averaged Lagrangian technique (given in Chapter V), as exemplified by the important paper of Simmons (1969). These interaction equations have two independent integrals. From the first two equations,

$$C_1^{-1}\dot{a}_1 a_1{}^* = -ia_1{}^* a_2{}^* a_3{}^* = C_2^{-1}\dot{a}_2 a_2{}^*$$

so that

$$\frac{d}{dt}\left(\frac{a_1 a_1^*}{C_1} - \frac{a_2 a_2^*}{C_2}\right) = 0$$

and

$$\frac{a_1 a_1^* - A^2(0)}{C_1} = \frac{a_2 a_2^* - B^2(0)}{C_2} = \frac{a_3 a_3^* - C^2(0)}{C_3}, \tag{11}$$

where $A^2(0)$ represents the initial value of $a_1 a_1^*$ and so on. The general solution can be expressed in the form of Jacobi elliptic functions, as shown by McGoldrick (1965) and Bretherton (1964). However, a number of interesting and important properties of the solution can be seen without this complexity.

The integrals (11) can be regarded as describing the partition of energy among the three components involved in the interaction. The quantities $a_1 a_1^*$, etc., are proportional to the energy density of the wave motion, and Bretherton has shown that if the property p is such that the constant of proportionality is purely numerical, then the interaction coefficients are proportional to the frequencies ω, which satisfy the resonance condition (9). Since (9) implies

$$\omega_1 + \omega_2 + \omega_3 = 0,$$

at least one of the frequencies, defined in this way, must be negative. Therefore at least one of the coefficients C_1, C_2, C_3 must be negative also. Suppose that it is C_1. If the energy density in component 1 increases at some instant, the combined energy density of the other two components must decrease. The solutions for the wave amplitudes are generally periodic, the period being ϵ^{-1} times a characteristic wave period. The total energy density in the three wave components is, of course, conserved; the changing partition among them is illustrated in Figure VII.3. In the example illustrated, the initial amplitude of the 1-component is zero, so that from (10)

$$\dot{a}_2 = \dot{a}_3 = 0 \qquad \text{at } t = 0,$$

and the initial growth of the 1-component is linear,

$$i a_1 = C_1 a_2^* a_3^* t,$$

and the energy density initially grows quadratically,

$$a_1 a_1^* = C_1^2 a_2 a_2^* a_3 a_3^* t^2.$$

From these initial conditions, the energy density in component 1 grows at the expense of the other two until one or other of these components is exhausted. At this point, the energy density in component 1 has ceased to grow; it then begins to decrease as the exhausted component is invigorated and the energy distribution returns to its initial partition.

Another interesting consequence of these equations is that if any wavetrain is capable of undergoing resonant interactions of this kind with nonzero interaction coefficients, it is intrinsically unstable to perturbations at a pair of wavenumbers such that the three constitute a resonant triad. This was pointed out by Hasselmann (1967).

Examples of second-order interactions have been studied in a variety of contexts. McGoldrick (1970) has shown that a purely sinusoidal gravity-capillary wave of wavenumber $(g\rho/2T)^{1/2}$ of any amplitude cannot propagate indefinitely on an inviscid liquid, but transfers energy

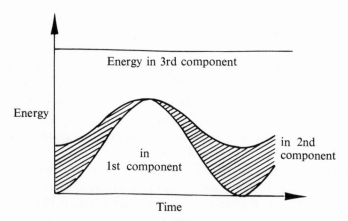

VII.3. A schematic representation of the changing energy partition among three wavenumbers in tertiary wave interactions. Initially the energy is confined to the 2 and 3 components; as the energy in the 1 component grows. it drains from the other two until the 2 component is exhausted. In the absence of dissipation or of interaction with further wavenumbers, the process is cyclic.

to its own second harmonic. The solution to the linear problem represents a theoretically possible state of dynamical equilibrium to the lowest order, which is in fact unstable at the next order of approximation. This effect arises as a consequence of the minimum in the phase velocity that occurs for gravity-capillary waves; for this particular wavenumber, the second harmonic of the primary wave corresponds in fact to a free wave mode traveling at the same speed with twice the wavenumber of the primary wave. The instability of this wave was demonstrated in a series of very clear experiments showing the decay of the primary wave and the initial amplification of the second harmonic followed by its decay under the influence of molecular viscosity. Phillips (1968) has considered the propagation of internal gravity waves in a uniformly stratified fluid in which there are weak periodic vertical variations in the horizontal current field. These can be considered as a limiting case of internal gravity waves as the wave frequency approaches zero and the wavenumber direction approaches the vertical. If this wavenumber is designated by $\mathbf{k}_1$, then wavenumbers $\mathbf{k}_2$ and $\mathbf{k}_3$, forming an isosceles triangle with $\mathbf{k}_1$ as shown in Figure VII.4, satisfy the resonance condition, with

$$\mathbf{k}_1 + \mathbf{k}_2 = \mathbf{k}_3, \qquad \omega_1 + \omega_2 = \omega_3,$$

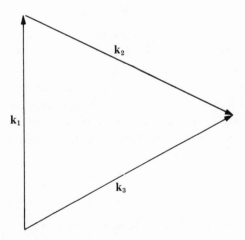

VII.4. A resonant interaction triad for internal gravity waves.

since $\omega_1 = 0$ and the frequencies of the 2, 3 components are equal because of their equal inclinations with the vertical. In this case, however, it is found that the interaction coefficients are such that $C_1 = 0$ and $C_2 = -C_3$. There is no energy exchange at all with the steady motion; it is, in a sense, catalytic, giving rise to exchanges between the inclined components. Such a current distribution, therefore, scatters an internal wave of this kind while not itself being affected.

Another example of great oceanographic interest has been studied experimentally; a large tank with a diffuse thermocline is capable of supporting may different modes of internal gravity waves, and Martin, Simmons, and Wunsch (1972) showed that among these modes there are many triads for which the resonance conditions are very nearly satisfied.

3. Interactions among Surface Gravity Waves

It appeared earlier that among pure gravity waves, no second-order resonant interactions are possible. The question then arises: Do they occur at a higher order?

In surface waves, the nonlinear terms arise as the result of Taylor series expansions of the free surface conditions, and they contain terms of all orders. In symbolic terms:

$$\phi_{tt} + g\phi_z = [\quad]\phi^2 + [\quad]\phi^3 + \cdots.$$

The quadratic terms give rise to simple sums and differences of both wavenumber and frequency, neither of which provide resonant contributions. The cubic terms, on the other hand, lead to forcing functions involving triple sums and differences of the wavenumbers and frequencies, and if these coincide with a natural (wavenumber, frequency) pair appropriate to free surface waves, then resonance is possible at this order. In other words, the third-order resonance conditions are of the form

$$\mathbf{k}_1 \pm \mathbf{k}_2 \pm \mathbf{k}_3 \pm \mathbf{k}_4 = 0,$$

$$\omega_1 \pm \omega_2 \pm \omega_3 \pm \omega_4 = 0.$$

It is possible to show by a slight modification of the earlier graphical procedure that solutions to this set of equations do exist. In Figure

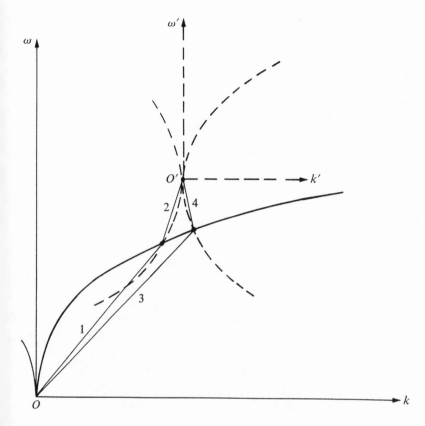

VII.5. A diagram in the wavenumber-frequency plane illustrating a quartet of interacting wave numbers in surface gravity waves.

VII.5, the origin O is such that the vector drawn to any point on the dispersion curve represents a wavenumber frequency pair for free surface waves. If we choose a second origin O', which does not lie on this curve, and construct the dispersion curves relative to this origin, we obtain a set of curved lines that intersect the original dispersion curve at two or more points. Any vectors from this new origin to the new dispersion curves also represent wavenumber, frequency pairs of free surface waves; and if we join the two intersection points to the origins O and O', we obtain a set of four vectors, forming a closed quadrilateral, whose wavenumbers and frequencies satisfy the resonance condition. In the example shown,

$$\mathbf{k}_1 + \mathbf{k}_2 = \mathbf{k}_3 + \mathbf{k}_4,$$

$$\omega_1 + \omega_2 = \omega_3 + \omega_4. \tag{12}$$

This diagram (Figure VII.5) in two dimensions represents a section of what is really a surface, axisymmetric about the ω axis. The frequency of free surface waves depends on the magnitude of the wavenumber vector $\mathbf{k}$ but not on its direction. Consequently, Figure VII.5 represents a section through two intersecting trumpet-shaped surfaces, and the complete set of interacting wavenumbers is represented by the locus of possible intersections. This is shown more economically in Figure VII.6, which represents the projection of intersection loci onto

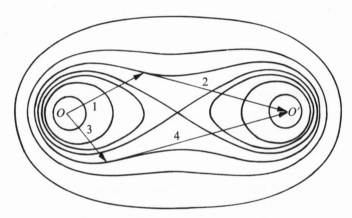

VII.6. The interaction diagram in the wavenumber plane for surface gravity waves. Any two points on the same member of the family of curves specify four wavenumbers capable of resonant interaction.

the wavenumber plane. The trumpet whose origin is at O is fixed; for each choice of O' sufficiently close to the dispersion curve, the intersection locus is a pair of loops, surrounding the projections of O and O' on the wavenumber plane. As the origin O' moves vertically, the projection of the intersection loci expands, becomes a figure eight when the two surfaces are tangential at the point midway between O

and O', then opens out into a single loop. The family of curves shown in Figure VII.6 is such that if we choose any two points lying on one family, then the wavenumbers $\mathbf{k}_1$, $\mathbf{k}_2$, $\mathbf{k}_3$, and $\mathbf{k}_4$ joining these points to O and O' form a resonant quartet that is capable of exchanging energy among its components but not with components outside. Again, the interactions are selective; not every set of four wavenumbers forming a closed quadrilateral is capable of resonant energy exchange. Such exchange is limited to those specified by this family of curves.

The interaction equations for these third-order resonances are rather more complicated than the simple second-order equations. They can be found in precisely the same way, however, so that

$$
\begin{aligned}
\omega_1\dot{a}_1 &= ia_1(g_{11}a_1a_1{}^* + g_{12}a_2a_2{}^* + g_{13}a_3a_3{}^* + g_{14}a_4a_4{}^*) + iha_2{}^*a_3a_4, \\
\omega_2\dot{a}_2 &= ia_2(g_{21}a_1a_1{}^* + \cdots \qquad\qquad\qquad\quad) + iha_1{}^*a_3a_4, \\
\omega_3\dot{a}_3 &= ia_3(g_{31}a_1a_1{}^* + \cdots \qquad\qquad\qquad\quad) + iha_1a_2a_4{}^*, \\
\omega_4\dot{a}_4 &= ia_4(g_{41}a_1a_1{}^* + \cdots \qquad\qquad\qquad\quad) + iha_1a_2a_3{}^*.
\end{aligned}
\tag{13}
$$

Equations of this type were first derived by Benney (1962). The terms on the righthand side are of two types. The first terms, involving the matrix g_{ij} (whose elements are *real* functions of the configuration of wavenumbers only), are associated with phase changes that modify but do not destroy the energy interchange. The terms inside the parentheses are all real; so the combination represents a rate of change of the complex amplitudes a_i which are always 90° out of phase from a_i; they therefore represent a turning in time of the complex vector, while its amplitude remains constant. In other words, these terms represent a modification to the wave frequency that results from the interaction; equivalently, they represent an effect of the wave amplitude on the infinitesimal dispersion relation. The diagonal terms describe the interaction of a wave with itself to the third order, with the increase in phase velocity of $\frac{1}{2}(k_ia_i)^2(g/k_i)^{1/2}$ (no summation) that was discovered by Stokes in 1847. The off-diagonal terms describe changes in the phase velocities that result from the mutual interaction of pairs of waves, an effect noted by Longuet-Higgins and Phillips (1962). The last terms in (13) describe the energy transfer among the different components. The real coefficient h that occurs in all of the equations is a complicated algebraic function of the wavenumbers of the order $k_1k_2k_3k_4$.

Equations (13) admit a number of simple integrals; there are three partition integrals:

$$\omega_1a_1a_1{}^* - \omega_2a_2a_2{}^* = const.,$$
$$\omega_3a_3a_3{}^* - \omega_4a_4a_4{}^* = const., \qquad (14)$$
$$\omega_1a_1a_1{}^* + \omega_3a_3a_3{}^* = const.$$

Since the energy density of a mode is $E_i = \frac{1}{2}\rho g a_i a_i^*$ (no summation), these integrals specify the sharing of the total energy among the four modes. The total energy of the four modes is a constant, as may be seen from (14) and the resonance conditions (12). Energy is gained by one pair at the expense of the other. If, at a certain instant, energy in component 1 is increasing, then it is also increasing in component 2 at the expense of 3 and 4. In this cubic interaction, the interaction time is of the order of the (wave slope)$^{-2}$ times a characteristic wave period; the interactions are not only selective but very weak.

Nonetheless, the absence of significant viscous dissipation makes the prospect of observing these interactions in the laboratory attractive. Two sets of experiments have been performed, one by Longuet-Higgins and Smith (1966) and the other, following the suggestion of Longuet-Higgins (1962) by McGoldrick, Phillips, Huang, and Hodgson (1966).

The most convenient experimental configuration, associated with the figure-eight loop of the family illustrated in Figure VII.6, is shown in Figure VII.7. Orthogonal wavenumbers $\mathbf{k}_1$ amd $\mathbf{k}_2$ are generated

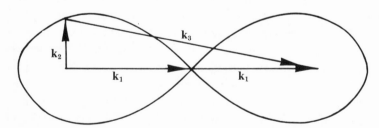

VII.7. The resonance loop for triads of interacting gravity waves in deep water.

mechanically by plungers on two adjacent sides of a rectangular tank (Figure VII.8). The wavenumber k_1 interacts with itself and with the wavenumber k_2 to generate the wavenumber $k_3 = 2k_1 - k_2$. If the frequency ratio of the primary wave is such that k_2 lies on the figure-eight curve, then the resonance condition is satisfied and the wavenumber k_3 grows and propagates across the tank to be absorbed at the beaches.

For this type of interaction, in which the wavenumber k_1 enters twice, the general interaction equations are of the form

$$\omega_1 \dot{a}_1 = ia_1(g_{11}a_1a_1{}^* + g_{12}a_2a_2{}^* + g_{13}a_3a_3{}^*) + iha_1{}^*a_2a_3,$$

$$\omega_2 \dot{a}_2 = ia_2(g_{21}a_1a_1{}^* + \cdots \qquad\qquad) + \tfrac{1}{2}iha_1{}^2a_3{}^*, \qquad (15)$$

$$\omega_3 \dot{a}_3 = ia_3(g_{31}a_1a_1{}^* + \cdots \qquad\qquad) + \tfrac{1}{2}iha_1{}^2a_2{}^*.$$

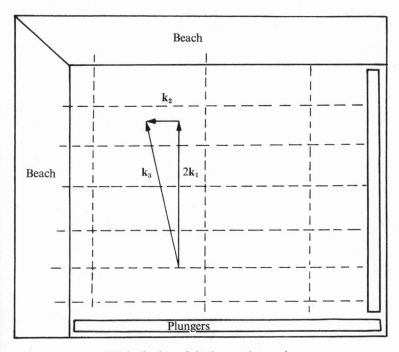

VII.8. A plan of the interaction tank.

The algebraic details are given by McGoldrick et al. (1966). Two partition integrals follow immediately:

$$\omega_1 a_1 a_1{}^* + 2\omega_2 a_2 a_2{}^* = const., \tag{16}$$

$$\omega_1 a_1 a_1{}^* + 2\omega_3 a_3 a_3{}^* = const.$$

From these it follows that any growth in the component with wave-number k_3 is accompanied by growth in the component with wave-number k_2, both at the expense of the k_1 component.

In the experiment, the time available for the interaction is only a few wave periods, very much less than the interaction time (ϵ^{-2}) times the wave period, so that the amplitude a_3 of the third component is always small compared with those of the other two. In this circumstance, the last of the set (15) reduces to

$$\omega_3 \dot{a}_3 = \tfrac{1}{2} i h a_1{}^2 a_2{}^*, \tag{17}$$

whence, since both a_1 and a_2 are, in effect, constant over this short time interval,

$$a_3 = \frac{i h a_1{}^2 a_2{}^*}{2\omega_3}\, t. \tag{18a}$$

Under these conditions, then, the amplitude of the third component is a linear function of the time over which the interaction has taken place. In the experiment, this is proportional to the distance d that the energy in this third wave component travels during the course of the interaction, and the solution can be expressed alternatively as

$$a_3 = (a_1 k_1)^2 (a_2 k_2) G(r,\ \theta) d, \tag{18b}$$

as shown by Longuet-Higgins and Smith (1966), where the interaction coefficient G is a function of the frequency ratio of the primary waves and their angle of intersection. In the particular configuration considered experimentally, the frequency ratio $r = 1.736$, and $G = 0.442$.

These expressions specify the growth of the third wave component when the resonance conditions are satisfied exactly. As in all resonance phenomena, however, lying near the resonant point is a band of wavenumbers that also are excited. The width of this band decreases as the interaction distance increases; the resonance becomes more sharply tuned. For wavenumbers near but not exactly at resonance, we have

$$a_3 = (a_1 k_1)^2 (a_2 k_2) G(r, \theta) d \left| \frac{\sin \delta k d}{\delta k d} \right|,$$

where δk is a measure of the mismatch in the wavenumber plane from resonance. In the configuration studied experimentally, $\delta k = 0.249 k_3 \delta r$. The detailed derivation of this expression is given by Longuet-Higgins and Smith (1966).

Some of the results of the experiments are shown in the accompanying figures. Figure VII.9 shows the spectrum of waves in the tank when the frequency ratio of the primary wave was adjusted for resonance. The two largest components are clearly the primary waves. Next, however, is the resonant and growing tertiary wave generated by the interaction of the two primary waves. The amplitude of this component is larger than that of any other produced by the interaction, even though this is a third-order effect and the interaction distance is short. A similar set of measurements made by McGoldrick et al. is shown in Figure VII.10. The shapes of the curves around the peaks correspond closely to the response shape of the filters used; the spectra are consistent with the line spectra shown in the diagram. At the top of Figure VII.10 the frequency ratio is adjusted for resonance, and again we observe that the component with frequency $2f_1 - f_2$ is comparable to or larger than that of any of the second-order components generated by the interaction. When the resonance is detuned, however, as in the lower part of Figure VII.10, this is no longer so; the third-order component recedes into the background.

Figure VII.11 is representative of the results obtained when the amplitude of the third component is measured as the frequency ratio f_1/f_2 passes through resonance. The width of the resonance band can be seen; it grows narrower as the interaction distance increases. The displacement of the resonant frequency ratio from the theoretical value of 1.736 is a consequence of amplitude dispersion—the fact that the wave frequency at a given wavenumber increases slightly with increasing wave slope. The experiments of Longuet-Higgins and Smith (1966) were conducted at a substantially larger wave slope than those of

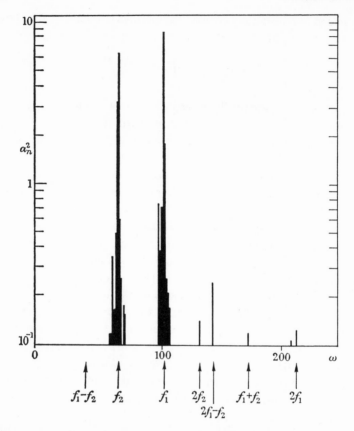

VII.9. The spectrum of waves in the tank in the measurements of Longuet-Higgins and Smith (1966). The squares of the Fourier coefficients are shown as a function of the harmonic number ω. The frequency ratio f_1/f_2 is adjusted for resonance. (From Phillips, 1967, by permission of The Royal Society, London.)

McGoldrick et al. (1966), and the shift in resonant frequency ratio is substantially larger. In Figure VII.12 the amplitude of the third component, suitably normalized, is plotted against the interaction distance. In both sets of experiments the amplitude evidently grows linearly, as the theory predicts, and the slope of the observed points is quite close to that calculated from the theory.

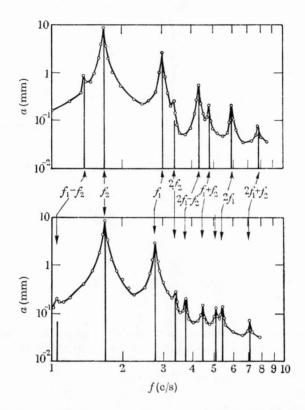

VII.10 The spectra of waves in the tank, measured by McGoldrick et al. (1966) both at resonance (above) and off resonance (below). (From Phillips, 1967, by permission of The Royal Society, London.)

Perhaps the most dramatic manifestation of these surface wave interactions is found in the instability of a Stokes wave, an effect discovered by Benjamin and Feir (1967). They found that when a classical Stokes wave suffers an infinitesimal perturbation at an adjacent wavenumber, the amplitude of the perturbation grows until the originally almost uniform wavetrain degenerates into a series of wave groups. The effect had long been observed in wave tanks; it is one of the problems in attempting to generate a steady wavetrain in a long tank. Benjamin and Feir's achievement is in demonstrating that this is a

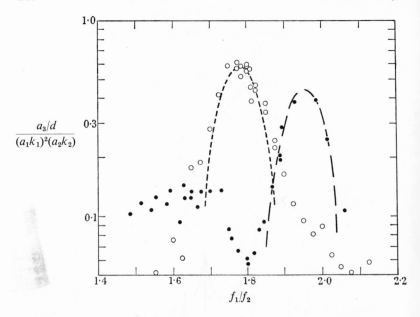

VII.11. The interaction products at a distance of 198 centimeters in the interaction tank. The open circles represent measurements by McGoldrick et al. (1966) and the solid circles by Longuet-Higgins and Smith (1966). The broken lines represent the theoretical band shapes. Note the shift in the resonant frequency ratio observed by Longuet-Higgins and Smith, resulting from the amplitude dispersion effect. (From Phillips, 1967, by permission of The Royal Society, London.)

genuine hydrodynamic instability and not (as had been suspected) a consequence of imperfections in the engineering design.

The phenomenon is closely related to the intrinsic instability pointed out by Hasselmann (1967), described earlier, with the additional property that amplitude dispersion effects are important. Its existence, which can be demonstrated from the equations given above, is proven by a different method in Chapter V, Section 7. When a_2 and a_3 are comparable, but both initially small compared with a_1, which is in effect constant, then to first order in the perturbation amplitudes the last of (15) reduces to

$$\omega_3 \dot{a}_3 = ia_3(g_{31}a_1a_1{}^*) + \tfrac{1}{2}iha_1a_2{}^* \tag{19}$$

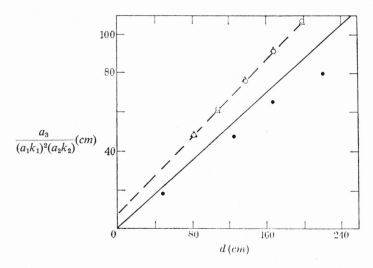

VII.12. The growth of amplitude of the interaction product with distance d. The continuous line represents the predicted rate, $A = 0.442d$. The solid circles represent measurements by Longuet-Higgins and Smith. The open circles and triangles represent two different series of measurements by McGold-rick et al. in which the slopes of the primary waves were also varied. (From Phillips, 1967, by permission of The Royal Society, London.)

with a similar equation for a_2. If

$$\beta_3 = a_3 \exp \left(\frac{-ig_{31}a_1a_1^*t}{\omega_3} \right),$$

$$\beta_2 = a_2 \exp \left(\frac{-ig_{21}a_1a_1^*t}{\omega_2} \right),$$

then

$$\omega_3\beta_3 = \tfrac{1}{2}iha_1^2\beta_2^* \exp \left[-i \left(\frac{g_{21}}{\omega_2} + \frac{g_{31}}{\omega_3} \right) a_1a_1^*t \right],$$

$$\omega_2\beta_2 = \tfrac{1}{2}iha_1^2\beta_3^* \exp \left[-i \left(\frac{g_{21}}{\omega_2} + \frac{g_{31}}{\omega} \right) a_1a_1^*t \right].$$

(20)

If, in addition, there is a slight mismatch from the resonance condition appropriate to infinitesimal waves to compensate for the modification in the dispersion relation associated with the first term in (19) (the amplitude dispersion), then a further factor exp $i\epsilon t$ is required on the right [Bretherton (1964), Phillips (1967)]; and, provided that

$$\epsilon = \left(\frac{g_{21}}{\omega_2} + \frac{g_{31}}{\omega_3}\right) a_1 a_1^*,$$

then the phase variation in (20) is completely annulled and reduces it to (17). We can express these resulting equations in terms of the (real) wave amplitudes α, by identifying (18a) and (18b); and they reduce to

$$\dot{\alpha}_3 = (\alpha_1 k_1)^2 \left(\frac{g}{\omega_3}\right) G(\alpha_2 k_2),$$

$$\dot{\alpha}_2 = (\alpha_1 k_1)^2 \left(\frac{g}{\omega_2}\right) G(\alpha_3 k_3).$$

These combined effects of resonant detuning and amplitude dispersion are the essential ingredients of the Benjamin-Feir instability of the Stokes wave, and, as in their analysis, a coupled set of linear equations for the perturbation amplitudes α_2 and α_3 have exponentially growing solutions

$$\alpha_2 = \omega_3^{3/2} C \exp \gamma t,$$

$$\alpha_3 = \omega_2^{3/2} C \exp \gamma t,$$

where $\gamma = G(\alpha_1 k_1)^2 (\omega_2 \omega_3)^{1/2}$. Both perturbations draw energy from the primary wavetrain, as the partition integrals (16) indicate. The Stokes wave is unstable to pairs of perturbation wavenumbers defined by points just inside the figure-eight loop, as illustrated in Figure VII.13. The growth rate is proportional to the square of the primary wave slope, and so is the bandwidth. The steeper the primary Stokes wave, the more rapid the instability and the wider the bandwidth over which it is found. In Benjamin and Feir's experiments, the perturbation wavenumbers were collinear with the primary wave; for these wavenumbers there is a zone of unstable wavenumbers lying on either side of the primary wave.

There is an irony here. In the early years of this century a vigorous controversy raged between Rayleigh and Burnside as to whether truly

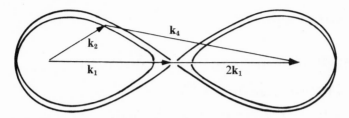

VII.13. A sketch of the zone of wavenumbers to which a Stokes wave with wavenumber k_1 is unstable. Any triangle whose vertex lies within the strips indicated specifies a pair of unstable wavenumbers. The experiments of Benjamin and Feir refer to the two-dimensional case in which the vertex lies in the direction of k_1.

steady-state solutions existed that represented a finite-amplitude surface wavetrain. It was finally proved by Levi-Civita in 1921 that such solutions did exist; now it turns out that they are unstable! [See Lamb (1932), p. 420.]

CHAPTER VIII

The Korteweg-deVries Equation: A Model Equation for Nonlinear Dispersive Waves

Robert M. Miura

One of the simplest model equations in the study of nonlinear dispersive wave phenomena is the Korteweg-deVries (K-dV) equation

$$u_t + \alpha u u_x + \beta u_{xxx} = 0, \tag{1}$$

just as the Burgers equation

$$u_t + u u_x = \nu u_{xx} \tag{2}$$

is the simplest model equation for nonlinear dissipative wave phenomena. The K-dV equation is a nonlinear partial differential equation with nonlinear term $\alpha u u_x$ and dispersive term βu_{xxx}. Physically, it describes the long-time evolution of small-but-finite amplitude nonlinear waves. The derivation of the equation for long-wavelength water waves is due to Korteweg and deVries (1895) and is sketched in Chapter IV. Other applications of the K-dV equation include magnetohydrodynamic waves in a cold plasma (Gardner and Morikawa, 1960); longitudinal vibrations of an anharmonic discrete-mass string (Zabusky, 1963, 1967, 1969; Kruskal, 1965); ion-acoustic waves in a cold plasma (Washimi and Taniuti, 1966); pressure waves in liquid-gas bubble mixtures (Wijngaarden, 1968); rotating flow down a tube (Leibovich, 1970); and longitudinal dispersive waves in elastic rods (Nariboli,

This work was supported by the Magneto-Fluid Dynamics Division, Courant Institute of Mathematical Sciences, New York University, under Contract AFOSR-71-2053 with the Air Force Office of Scientific Research.

1969). The derivation of the K-dV equation (as well as the Burgers equation) in a general class of problems has been given by Su and Gardner (1969) and Taniuti and Wei (1968). Zabusky and Galvin (1971) have shown that the K-dV equation gives an accurate description of slightly dissipative water waves. In addition to these specific applications, the K-dV equation serves as a model on which to test new techniques for solving nonlinear equations and for determining properties of such equations and their solutions.

Although in the applications, α and β in (1) take on specific values, a simple (real) scaling of the variables permits choosing α and β arbitrarily. Most mathematical treatments of the K-dV equation choose $\alpha = 1$ and $\beta = 1$. For most of the present chapter it is convenient to let $\alpha = -6$ and $\beta = 1$; then the K-dV equation has the form

$$Pu \equiv u_t - 6uu_x + u_{xxx} = 0. \tag{3}$$

Although the K-dV equation arises in numerous applications, recent mathematical interest in it has been due to its various properties and solutions. The equation possesses steady progressing wave solutions in the form of uniform wavetrains and solitary waves. The uniform wavetrain solutions, called cnoidal waves, are periodic and may be expressed in terms of Jacobi elliptic functions. On the other hand, solitary waves have the form

$$u(x, t) = -\tfrac{1}{2}a^2 \operatorname{sech}^2 \left[\tfrac{1}{2}a(x - a^2 t)\right]. \tag{4}$$

[Note that the solitary waves are negative because the coefficient of the nonlinear term in (3) is negative.] This means that solitary waves propagate to the right at a speed a^2, which is proportional to the amplitude, and have a width that is inversely proportional to the square root of the amplitude.

One of the most interesting properties of the solutions is the "linear-like" behavior of the solitary waves. It is well known that solutions of linear equations can be superposed to yield new solutions. One consequence is that two linear steady progressing waves can pass through each other and emerge unchanged in form. In general, superposition of solutions to nonlinear equations fails to yield new solutions. However, in numerical calculations by Zabusky and Kruskal (1965) it was observed that two distinct solitary waves, i.e. with distinct amplitudes,

interact nonlinearly but emerge from the interaction unchanged. This
resemblance of solitary waves to particles led to the name "solitons."
Specifically, consider two well-separated distinct solitons, with the
"taller" soliton to the left of the "shorter" one [Figure VIII.1(a)].
Because the taller soliton has higher speed, it will overtake the shorter
(and hence slower) soliton; they will interact nonlinearly according to
the K-dV equation, and surprisingly they emerge from the interaction
unchanged in form [Figure VIII.1(b)]. These (nonlinear) solitons can be

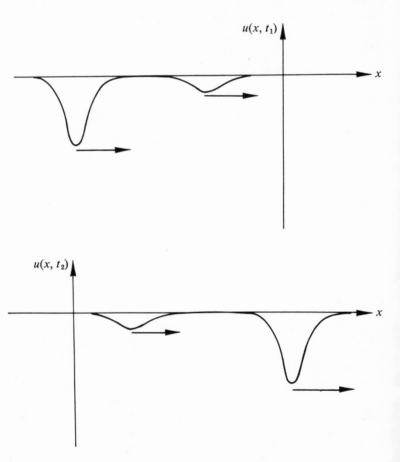

VIII.1. Interaction of solitons ($t_2 > t_1$).

distinguished from linear solutions by observing that they are shifted in their positions relative to where they would have been if no interaction had occurred. This behavior for two solitons was verified numerically by Zabusky (1967) and was proved by Lax (1968). A method will be presented below for solving the K-dV equation that can be used to prove the corresponding result for N distinct solitons, where N is an arbitrary but finite number.

More generally, it has been shown numerically (Zabusky, 1968) that in addition to (or instead of) solitons propagating to the right, some solutions will have a dispersing oscillatory state propagating to the left with solitons propagating to the right (Figure VIII.2). The propagation of the oscillatory state to the left is clear because of the negative group velocity of the linear waves.

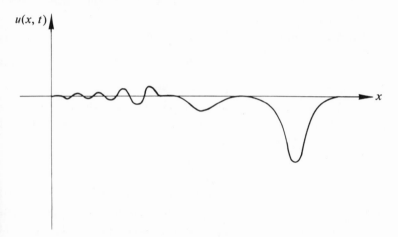

VIII.2. Solution with both solitons and an oscillatory state.

1. Conservation Laws and a Nonlinear Transformation

The K-dV equation possesses many conservation laws—that is, equations in the form

$$T_t + X_x = 0, \tag{5}$$

where T is the *conserved density* and $-X$ is called the *flux*. Our concern is primarily with conserved densities and fluxes which are polynomials in u and its x derivatives. For example, the K-dV equation can immediately be rewritten in conservation form

$$u_t + (-3u^2 + u_{xx})_x = 0.$$

Multiplying the K-dV equation by u yields another conservation law:

$$(\tfrac{1}{2}u^2)_t + (-2u^3 + uu_{xx} - \tfrac{1}{2}u_x{}^2)_x = 0.$$

A third conservation law was found by Whitham (1965a), and two more were found by Kruskal and Zabusky (Zabusky, 1967). In all, polynomial conserved densities for eleven conservation laws were obtained explicitly (Miura et al., 1968; Kruskal et al., 1970), and it was conjectured that an infinite number of them existed. An early attempt to prove this conjecture resulted in an explicit formula for the conserved densities if they existed (Kruskal et al., 1970), but this formula could not be used to prove existence.

Another equation similar to the K-dV equation,

$$Qv \equiv v_t - 6v^2v_x + v_{xxx} = 0, \tag{6}$$

also possesses many conservation laws (Miura et al., 1968). [The corresponding equation with a positive sign before the nonlinear term arises in the theory of anharmonic discrete-mass strings (Zabusky, 1967).] However, if one considers the generalization

$$w_t - 6w^p w_x + w_{xxx} = 0,$$

where $p \geq 3$ is an integer, there are only three polynomial conserved densities. (General results for a class of equations that includes all of the equations above will be given below.) This led to the conjecture that perhaps the solutions of the K-dV equation and the v equation (6) were related. To find a relationship by direct comparison of (3) and (6) seemed hopeless, but a comparison between conservation laws led to many conditions from which the desired relation follows (Miura, 1968).

THEOREM 1. *If v satisfies (6), $Qv = 0$, then u defined by*

$$u \equiv v^2 + v_x \tag{7}$$

satisfies the K-dV equation, $Pu = 0$.

Proof.
$$Pu = \left(\frac{\partial}{\partial x} + 2v\right) Qv. \tag{8}$$

Note that the converse of this result is not true because of the additional operator on Qv. The transformation (7) has led to many important results, as we shall see below.

We return to the question of the existence of polynomial conservation laws for the K-dV equation. Gardner (Miura, 1968) generalized the transformation (7) by introducing a formal parameter ϵ. This may be accomplished as follows. The K-dV equation is galilean invariant: in other words, the change of variables

$$t \to t', \quad x \to x' - 6ct', \quad u \to u' + c,$$

leaves the form of the equation unchanged in the new primed variables; on the other hand, the v equation is not galilean invariant. Introduce the following change of variables

$$t' \equiv t, \quad x' \equiv x + \frac{3}{2\epsilon^2} t, \quad u(x, t) \equiv u'(x', t') + \frac{1}{4\epsilon^2},$$

$$v(x, t) \equiv \epsilon w(x', t') + \frac{1}{2\epsilon},$$

then the K-dV equation remains unchanged but the transformation (7) now becomes

$$u = w + \epsilon w_x + \epsilon^2 w^2, \tag{9}$$

where all primes have been dropped and the equations for u and w are related by (8):

$$0 = u_t - 6uu_x + u_{xxx} = \left(1 + \epsilon \frac{\partial}{\partial x} + 2\epsilon^2 w\right) [w_t - 6(w + \epsilon^2 w^2)w_x + w_{xxx}]$$

$$\equiv LRw. \tag{10}$$

The relations (9) and (10) are generalizations of (7) and (8), respectively. Since the K-dV equation does not contain ϵ, only w depends on ϵ. In fact, both (9) and (10) can be solved for w and Rw respectively as formal power series in ϵ. The formal series for w has the form

$$w = w_0 + \epsilon w_1 + \epsilon^2 w_2 + \cdots$$
$$= u - \epsilon u_x - \epsilon^2 (u^2 - u_{xx}) + \cdots, \tag{11}$$

where w_i, $i = 0$, 1, 2, ..., are polynomials in u, u_x, u_{xx}, ... Since $L = 1 + \epsilon M$,

$$Rw = w_t + (-3w^2 - 2\epsilon^2 w^3 + w_{xx})_x = 0 \tag{12}$$

(written in conservation form) to all powers in ϵ. Thus, not only is w a conserved density, but direct substitution into $Rw = 0$ and setting the coefficient of each power of ϵ equal to zero yields conservation laws with conserved densities w_0, w_1, ... Each of these is a distinct conservation law for the K-dV equation. One can prove, however, (Miura et al., 1968) that only the coefficients of even powers of ϵ lead to nontrivial ones.

It is obvious from (7) and (9) that each conservation law for the K-dV equation yields one for the v equation and one for the w equation. Hence, there is an infinite number of conservation laws for each of them.

Consider the general class of equations

$$\frac{\partial u}{\partial t} + u^p \frac{\partial u}{\partial x} + \frac{\partial^r u}{\partial x^r} = 0, \tag{13}$$

where $p = 0$, 1, 2, ... and $r = 1$, 2, 3, ... This class includes (with different coefficients) the K-dV equation, the v equation, and the Burgers equation as special cases. We obtain the following results for conservation laws of (13) [for details and proofs see Kruskal and Miura (to be published)]:

1. If r is even, then only one polynomial conservation law exists—namely, the equation itself.
2. If $r = 2q + 1$, $q = 0$, 1, 2, ..., then at least three conservation laws exist—namely, those with conserved densities

$$T_1 = u,$$
$$T_2 = u^2,$$
$$T_3 = u^{p+2} + \frac{1}{2}(-1)^q(p+1)(p+2)\left(\frac{\partial^q u}{\partial x^q}\right)^2. \tag{14}$$

Table VIII.1 lists the total number of distinct polynomial conservation laws that exist for the corresponding equations identified by the two integers p and q. Those equations with $p = 0$ are linear and are easily shown to have an infinite number of polynomial conservation

Table VIII.1

q \ p	0	1	2	$p \geq 3$
0	∞	∞	∞	∞
1	∞	∞	∞	3
$q \geq 2$	∞	3	3	3

laws, namely those having conserved densities $(\partial^n u/\partial x^n)^2$, $n = 0, 1, 2, \ldots$. For $q = 0$, a new variable $v \equiv u^p + 1$ can be defined satisfying the equation $v_t + v v_x = 0$, for which it is trivial to produce an infinite number of polynomial conservation laws (as well as other types). The two other equations possessing an infinite number of polynomial conservation laws are the K-dV equation and the v equation (6). The remaining equations in this class (13) possess only three polynomial conservation laws, and it is not understood at this time why the K-dV equation and the v equation are special or what prevents the other equations from having more polynomial conservation laws.

Another type of conservation law for the class of equations with $p = 1$ involves x and t explicitly as well as u and x derivatives of u,

$$\left(xu - \frac{1}{2}tu^2\right)_t + \left[\frac{1}{2}xu^2 + x\frac{\partial^{2q}u}{\partial x^{2q}} - \frac{\partial^{2q-1}u}{\partial x^{2q-1}}\right.$$
$$\left. - t\sum_{k=0}^{q-1}(-1)^k \frac{\partial^k u}{\partial x^k}\frac{\partial^{2q-k}u}{\partial x^{2q-k}} + \frac{1}{2}(-1)^{q+1}t\left(\frac{\partial^q u}{\partial x^q}\right)^2\right]_x = 0. \quad (15)$$

Integrating this on the interval $-\infty < x < \infty$ for solutions that vanish rapidly at infinity, one obtains

$$\frac{d}{dt}\int_{-\infty}^{\infty} xu \, dx = \frac{1}{2}\int_{-\infty}^{\infty} u^2 \, dx = const.,$$

which says that the first moment of the solution is a linearly increasing function of time—that is, the "center of mass" of the solution is moving steadily to the right.

One can obtain conservation laws using Noether's theorem (Gel'fand

and Fomin, 1963). The invariances leading to the three conserved densities (14) are translations in v, where $u \equiv v_x$, x (space), and t (time). The conservation law (15), which contains x and t explicitly in addition to u, u_x, u_{xx}, ..., arises from galilean invariance.

2. Method of Exact Solution

A goal of applied mathematics is to determine the exact solution of a problem described in terms of mathematical equations. In general, however, one usually falls far short of this goal and is content to obtain properties of the equations and their solutions by approximate and qualitative methods. Nevertheless, we achieve the goal for a special class of solutions of the K-dV equation by obtaining an exact linearization of the general initial-value problem with initial data $u(x, 0) = f(x)$, $-\infty < x < \infty$, where $f(x)$ vanishes rapidly as $|x| \to \infty$.

2a. Role of the Schrödinger Equation

The key to the method of solution is the transformation (7)

$$u = v^2 + v_x,$$

which is a Ricatti equation for v if we assume u is a known function. There is a standard linearizing transformation for the Ricatti equation given by

$$v \equiv \frac{\psi_x}{\psi},$$

so that (7) becomes

$$\psi_{xx} - u\psi = 0.$$

This is almost recognizable as a form of the one-dimensional, time-independent Schrödinger equation in quantum mechanics. However, a basic ingredient, the energy levels (eigenvalues), is missing. As pointed out earlier, the K-dV equation is galilean invariant, so u can be translated by a constant and x differentiations remain unchanged, whence

$$\psi_{xx} - (u - \lambda)\psi = 0, \tag{16}$$

where ψ is the wave function, u is the potential, and λ corresponds to the energy level. Note that the potential u depends on x and on a parameter t (not to be confused with the usual time variable in the Schrödinger problem). The variation of u with respect to t is, of course, still governed by the K-dV equation.

The usual problem in quantum mechanics is to determine the wave function and energy states for given potential u. The objective here is to determine u, not ψ and λ. This is called the "inverse scattering" problem in quantum mechanics.

2b. The Inverse Scattering Problem

The literature treating the inverse scattering problem is extensive, and the reader is referred to the papers of Gel'fand and Levitan (1955), Kay (1955), Kay and Moses (1956), and Levinson (1953), all of which bear directly on the present discussion.

The solution of the inverse scattering problem is reduced to the problem of solving a linear integral equation, the Gel'fand-Levitan equation,

$$K(x, y) + B(x + y) + \int_x^\infty B(y + z)K(x, z)\, dz = 0, \tag{17}$$

where B is a known function determined by certain properties of the wave functions and is discussed below. Note that this is a linear integral equation for $K(x, y)$, a function of two variables. If the Gel'fand-Levitan equation can be solved for $K(x, y)$, then the potential u is given by

$$u(x) = -2 \frac{d}{dx} K(x, x). \tag{18}$$

The complete specification of the Gel'fand-Levitan problem requires a detailed definition of the kernel and inhomogeneous term

$$B(x + y) = \frac{1}{2\pi} \int_{-\infty}^\infty b(k)e^{ik(x+y)}\, dk + \sum_{n=1}^N c_n^2 e^{-\kappa_n(x+y)}, \tag{19}$$

where the integral represents contributions from the continuous spectrum of the Schrödinger operator and the sum is from the discrete spectrum

$$\lambda = k^2 \quad \text{and} \quad \lambda^{(n)} = -\kappa_n^2, \qquad n = 1, 2, \ldots, N. \tag{20}$$

More specifically, the quantities in B are defined as follows:

(i) $\lambda^{(n)}$—discrete eigenvalues. These are values of λ leading to eigenfunction solutions—solutions that vanish at infinity and are square integrable. Quantum mechanically, these discrete eigenvalues $(-\kappa_n{}^2)$ are simply the bound state energies. (Hereafter when the word "eigenvalue" is used, it may refer to $\lambda^{(n)}$ or to κ_n; in each instance the meaning will be clear.)

(ii) c_n—normalization constants. If the eigenfunction corresponding to κ_n is normalized

$$\int_{-\infty}^{\infty} \psi^{(n)^2} \, dx = 1, \tag{21}$$

then c_n is defined by the asymptotic behavior of $\psi^{(n)}$;

$$\psi^{(n)} \sim c_n \exp\left(-\kappa_n x\right) \qquad \text{as } x \to \infty. \tag{22}$$

(iii) $b(k)$—reflection coefficient. For the continuous spectrum, the wave function ψ is a linear combination of $\exp(\pm ikx)$, because u vanishes as $|x| \to \infty$. We impose the conditions

$$\psi \sim e^{-ikx} + b(k)e^{ikx}, \qquad x \to \infty, \tag{23}$$

and

$$\psi \sim a(k)e^{-ikx}, \qquad x \to -\infty. \tag{24}$$

Physically, the terms on the right correspond to a steady radiation of plane waves propagating into the potential from infinity; to an amount $b(k)$, called the reflection coefficient, being reflected from the potential; and to an amount $a(k)$, called the transmission coefficient, being transmitted through the potential. In particular,

$$|a|^2 + |b|^2 = 1.$$

Clearly, to solve the inverse scattering problem, namely to find u, the quantities κ_n, c_n, and $b(k)$ must be known. However, to determine these quantities we must know u. To escape from this vicious circle, a connection between the scattering problem and the K-dV equation will be established.

2c. The K-dV Equation and the Inverse Scattering Problem

The all-important link between the inverse scattering problem and the K-dV problem is the parametric dependence on t of the potential u. This dependence on t is governed by the evolution of the K-dV solutions. Also, ψ and λ must depend on t. To study this dependence on t, we derive evolution equations for ψ and λ by solving the eigenvalue equation for u and substituting into the K-dV equation. The result (after multiplying by ψ^2) is

$$\lambda_t \psi^2 + [\psi \mathscr{R}_x - \psi_x \mathscr{R}]_x = 0, \tag{25}$$

where

$$\mathscr{R} \equiv \psi_t + \psi_{xxx} - 3(u + \lambda)\psi_x.$$

THEOREM 2. *Consider the eigenvalue equation*

$$\psi_{xx} - (u - \lambda)\psi = 0, \qquad -\infty < x < \infty.$$

If u satisfies the K-dV equation and vanishes (rapidly) as $|x| \to \infty$, then the discrete eigenvalues $\lambda^{(1)}$, $\lambda^{(2)}$, ..., $\lambda^{(N)}$ are constants.

Proof. For an eigenvalue $\lambda^{(n)}$, the corresponding eigenfunction vanishes as $|x| \to \infty$; thus, integrating (25) over the infinite domain yields the condition

$$\lambda^{(n)}_t \int_{-\infty}^{\infty} \psi^{(n)^2} dx = 0,$$

which together with (21) gives $\lambda^{(n)} = const$. (This result is proved in Chapter XI by a method that is capable of extension to equations other than the K-dV equation.)

The usefulness of this result is evident—the discrete eigenvalues can be computed from the initial data for the K-dV equation and they will remain eigenvalues throughout the evolution of the solution. For the continuous spectrum (in other words, for positive values of λ) we can choose λ equal to a constant so that (25) can be integrated twice, yielding

$$\psi_t + \psi_{xxx} - 3(u + \lambda)\psi_x = C\psi + D\phi, \tag{26}$$

where C and D are integration "constants" at most dependent on t. The additional function ϕ is a solution linearly independent of ψ and is given by

$$\phi = \psi \int^x \frac{dx}{\psi^2}.$$

To completely define the inverse scattering problem, we determine the remaining quantities.

THEOREM 3. *If u vanishes sufficiently rapidly as* $|x| \to \infty$, *then*

$$c_n(t) = c_n(0)e^{4\kappa_n^3 t}, \tag{27}$$

$$b(k, t) = b(k, 0)e^{8ik^3 t}, \tag{28}$$

and

$$a(k, t) = a(k, 0), \tag{29}$$

where $c_n(0)$, $b(k, 0)$, *and* $a(k, 0)$ *are determined from the initial data for the K-dV equation.*

Proof. If ψ is an eigenfunction, then ϕ blows up exponentially as $x \to \infty$; thus $D \equiv 0$. Multiplying (26) by ψ and integrating over the infinite domain requires

$$\int (\tfrac{1}{2}\psi^2)_t \, dx + \int (\psi\psi_{xx} - 2\psi_x^2 - 3\lambda\psi^2)_x \, dx = C\int \psi^2 \, dx.$$

But the eigenfunctions are normalized so $C \equiv 0$. For $x \to \infty$, $\psi^{(n)}$ satisfies (22), so that (26) yields

$$c_{n_t} - 4\kappa_n^3 c_n = 0,$$

the solution being (27).

For the continuous spectrum, we consider a steady radiation of plane waves from $+\infty$ only and study the behavior of ψ as $x \to -\infty$. Using (24), (26) becomes

$$a_t + 4ik^3 a = Ca + \frac{D}{a}\int^x e^{2ikx} \, dx,$$

so $D \equiv 0$, leaving

$$a_t + (4ik^3 - C)a = 0.$$

For $x \to \infty$, ψ satisfies (23), and the vanishing of the coefficients of the linearly independent functions $\exp(\pm ikx)$ in (26) requires

$$C = 4ik^3 \quad \text{and} \quad b_t - 8ik^3 b = 0,$$

resulting in (28) and (29).

2d. Summary of the Method of Solution

A short summary of the method of solution at this point may help to crystallize an admittedly complex procedure. We wish to solve the initial-value problem

$$u_t - 6uu_x + u_{xxx} = 0, \quad -\infty < x < \infty, \quad t < 0,$$
$$u(x, 0) = f(x).$$

First, we solve the eigenvalue problem

$$\psi_{xx} - [f(x) - \lambda]\psi = 0, \tag{30}$$

from which κ_n, $c_n(0)$, and $b(k, 0)$ are determined. Then (27) and (28) yield the time-dependent quantities $c_n(t)$ and $b(k, t)$. These determine $B(x + y)$ explicitly as

$$B(x + y; t) = \frac{1}{2\pi} \int_{-\infty}^{\infty} b(k, 0)e^{i[8k^3 t + k(x+y)]} \, dk + \sum_{n=1}^{N} c_n^2(0)e^{8\kappa_n^3 t - \kappa_n(x+y)}, \tag{31}$$

so that the Gel'fand-Levitan equation is defined. If we can solve this integral equation, then the solution of the initial-value problem for the K-dV equation is simply

$$u(x, t) = -2 \frac{d}{dx} K(x, x; t), \tag{32}$$

where t in K is treated as a parameter.

One might argue that we have merely replaced one difficult problem with another one. However, two major simplifications have been achieved. First, the equations involved, (30) and the Gel'fand-Levitan equation, are linear; second, t enters the problem only parametrically. Moreover, κ_n and c_n are simply numbers, and $b(k)$ is a function to be used only under an integral.

2e. Soliton Solutions

There is a special class of initial conditions for which we can obtain exact solutions, namely those initial conditions that evolve exactly into only solitons. We call them *N-soliton solutions* (or simply soliton solutions), where N denotes the number of solitons. To demonstrate the method for obtaining these exact solutions we consider the solitary wave solution and a 2-soliton solution.

The solitary waves have the general form (4). For convenience, let $a = 2$; then as initial condition take

$$u(x, 0) = f(x) = -2 \operatorname{sech}^2 x.$$

The corresponding eigenvalue problem can be solved exactly (Landau and Lifshitz, 1958) with one discrete eigenvalue $\kappa_1 = 1$ and corresponding normalization constant $c_1(0) = \sqrt{2}$. Also $b(k, 0) = 0$; hence, $b(k, t)$ vanishes for all t, which means that a solitary wave is reflectionless—that is, one could not detect that the potential was there by impinging plane waves on it from infinity and measuring the reflections.

The Gel'fand-Levitan equation becomes

$$K(x, y; t) + 2e^{8t-x-y} + 2e^{8t-y} \int_x^\infty K(x, z; t) e^{-z} \, dz = 0.$$

An attempt to separate variables in the form $K(x, y; t) = L(x; t) \exp(-y)$ produces

$$L(x; t) + 2e^{8t-x} + e^{8t} L(x; t) \int_x^\infty e^{-2z} \, dz = 0,$$

which leads to the solution

$$K(x, x; t) = -\frac{2}{1 + e^{2x-8t}}.$$

Then the solution of the initial-value problem for the K-dV equation is

$$u(x, t) = -2 \frac{d}{dx} K(x, x; t) = -2 \operatorname{sech}^2 (x - 4t),$$

which is precisely (4) (with $a = 2$).

Next consider the initial condition

$$u(x, 0) = -6 \operatorname{sech}^2 x.$$

This does not lead to a solitary wave solution because the ratio of amplitude to width is not correct. Again the reflection coefficient is zero, but now there are two distinct eigenvalues, $\kappa_1 = 1$ and $\kappa_2 = 2$. The exact solution is

$$u(x, t) = -12 \frac{3 + 4 \cosh (2x - 8t) + \cosh (4x - 64t)}{[3 \cosh (x - 28t) + \cosh (3x - 36t)]^2}.$$

Asymptotically for $t \to \infty$, two solitons emerge having amplitudes 2 and 8.

It has been shown (Gardner et al., 1967 and another paper to be published) that whenever $b(k, t) = 0$, only solitons emerge from the initial condition—that is, the solution evolves exactly into N solitons. For these solutions the Gel'fand-Levitan equation is easily reduced to a finite system of linear algebraic equations that can always be solved explicitly. From this system of equations one sees that the N-soliton solutions can be written as rational functions of exponentials. A long-time asymptotic analysis of the exact solutions shows that there is a one-to-one correspondence between the eigenvalues and the solitons that emerge from the initial data; the solitons are uniquely characterized by a single parameter (aside from a phase shift). This correspondence is given by

$$-\kappa_n^2 = \lambda^{(n)} = \tfrac{1}{2}u_{\min}^{(n)} = -\tfrac{1}{4}a_n^2, \qquad n = 1, 2, \ldots, N, \tag{33}$$

where $u_{\min}^{(n)}$ denotes the minimum value of the nth soliton and a_n corresponds to the parameter appearing in (4) with a_n^2 equal to the speed.

Unfortunately, to date, no exact solutions have been obtained for more general initial data—that is, where the solution has both solitons and an oscillatory state in it (Figure VIII.2). For these cases the reflection coefficient is not identically zero, and the major mathematical difficulty arises from the integral contribution in $B(x + y)$.

3. A Nonlinear WKB Method of Solution

One form of the K-dV equation that is of interest physically and has received much study is

$$u_t + uu_x + \delta^2 u_{xxx} = 0, \tag{34}$$

where $\delta^2 \ll 1$ is a small parameter depending on various physical quantities. Computer-calculated solutions of this equation, on a finite interval, have been obtained for various values of δ^2. One practical problem in obtaining these solutions is that as δ^2 takes on smaller and smaller values, oscillations in the solutions become more and more closely spaced, and it is necessary to introduce additional space steps to accurately describe the solution. Owing to factors such as numerical accuracy and stability, these additions lead to a severe restriction on the total evolution time of the solution that can be computed. To try to overcome this difficulty, Kruskal and Zabusky (1963) (Zabusky, 1967) used the fact that the solutions become "nearly periodic" for small δ^2 and developed a method for treating such nonlinear waves. (By "nearly periodic" we mean that the changes in amplitude and wavelength are $O(\delta)$ over many oscillations in the solutions; these should not be confused with "almost periodic" functions). This method arises as a natural nonlinear generalization of the usual linear WKB method and is similar to the averaging method of Whitham (1965a, 1965b, 1967, 1970, Chapter V of this book) and Luke (1966). (This method has not been called an averaging method here; although averaging is an essential part, it is by no means all of it.) We will consider the basic ideas of the method and state a few results [see Miura and Kruskal (1973) for details].

As is well known, the linear WKB method obtains approximations to nearly periodic solutions of the linear ordinary differential equation

$$\delta^2\psi_{xx} + V(x)\psi = 0, \qquad \delta^2 \ll 1, \tag{35}$$

where $V(x)$ varies little over many oscillations in the solution ψ. [Any resemblance of (35) to (16) is purely coincidental!] The WKB method assumes a solution in the form

$$\psi \sim W(x)e^{i\theta}, \qquad \theta \equiv \frac{B(x)}{\delta}, \tag{36}$$

where W and B are to be determined. Having introduced two functions W and B in place of ψ, we have the freedom to specify one arbitrary condition between them.

We now generalize the ideas of the WKB method to treat nearly periodic solutions (in x) of a nonlinear partial differential equation. Because of the nonlinearity, we cannot expect the simple exponential form of (36) to hold for (34); therefore, we assume the formal series representation

$$u(x, t; \delta) \sim U(\theta, x, t; \delta) = U^{(0)}(\theta, x, t) + \delta U^{(1)}(\theta, x, t) + \cdots, \quad (37)$$

where $\theta = \theta(x, t; \delta)$. Here the dependence of U explicitly on x and t is analogous to the dependence of W on x, and the θ dependence generalizes the exponential in (36). One important unsolved problem is to establish in what sense the series representation $U(\theta, x, t; \delta)$ on the right of (37) approximates the actual solution $u(x, t; \delta)$ on the left.

The series representation (37) is very general, but we obtain major simplifications after we retain two properties from the WKB method— periodicity and one arbitrary condition. We assume that U is strictly periodic in θ—that is, θ is an "anglelike" variable, and since both U and θ have been introduced, we choose our free condition such that the period in θ is unity, namely

$$U(\theta, x, t; \delta) = U(\theta + 1, x, t; \delta). \quad (38)$$

The lack of a known dependence of U on θ is resolved by treating θ as an additional independent variable and later find its dependence on x and t. This has been called the "method of extension" by Sandri (1965) and is basically the method of multiple scales used in Chapter IV and in particular in Chapter V, Section 8. To illustrate, Figure VIII.3 shows a solution surface $U = U(\theta, x)$ in the two variables x and θ (for simplicity in this illustration the dependence of U on t is suppressed). One of the objectives of the method is to determine $\theta = \theta(x)$ so that the solution $u(x)$ to the original problem corresponds to the curve obtained by the intersection of the surfaces $U = U(\theta, x)$ and $\theta = \theta(x)$.

3a. Formulation

To apply the nonlinear WKB method to the K-dV equation, we first select the dependence of θ on δ such that to leading order there is minimal simplification in the complexity of the equation (Kruskal, 1963). For the K-dV equation all terms can be made equally important, and since the nonlinear and dispersive terms must then balance, we assume

$$\theta = \frac{B(x, t; \delta)}{\delta}, \quad (39)$$

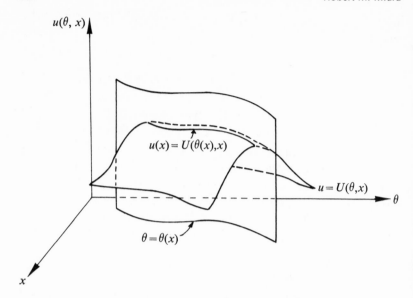

VIII.3. Determination of $u(x)$ from the extended solution.

where $B = O(1)$ is a formal power series in δ. Thus if B changes by a finite amount, θ will vary rapidly. Our objective, therefore, is to obtain $U(\theta, x, t; \delta)$ and $B(x, t; \delta)$ as formal power series in δ.

To complete the formulation of this extended problem, we derive the governing equation for U. The fact that only partial derivatives of θ will appear in this equation indicates the arbitrariness in the zero of θ. Therefore we define

$$L \equiv B_t, \quad K \equiv B_x, \quad l \equiv \frac{B_t}{B_x} = \frac{L}{K}, \tag{40}$$

which are independent of θ. The partial differential operators become

$$\frac{\partial}{\partial t} \to \frac{L}{\delta} \frac{\partial}{\partial \theta} + \frac{\partial}{\partial t}$$

and

$$\frac{\partial}{\partial x} \to \frac{K}{\delta} \frac{\partial}{\partial \theta} + \frac{\partial}{\partial x}.$$

Using these new operators in (34) and multiplying by δ/K yields the "extended" K-dV equation

$$lU_\theta + UU_\theta + K^2 U_{\theta\theta\theta} + \delta\left[\frac{1}{K}(U_t + UU_x) + 3(KU)_{x\theta\theta}\right]$$

$$+ \delta^2\left\{\frac{1}{K}\left[K_{xx}U + 3(KU_x)_x\right]\right\}_\theta + \delta^3 \frac{1}{K} U_{xxx} = 0. \qquad (41)$$

Substituting (37) into (41) and keeping only the leading-order terms, we have the ordinary differential equation in θ for $U^{(0)}$

$$l^{(0)}U^{(0)}_\theta + U^{(0)}U^{(0)}_\theta + K^{(0)^2}U^{(0)}_{\theta\theta\theta} = 0, \qquad (42)$$

where $l^{(0)}$ and $K^{(0)}$ are the leading-order terms in formal expansions of l and K.

It is important to point out that in the limit $\delta \to 0$, we obtain (42) and not an equation in the form $u_t + uu_x = 0$ as one does in treating the Burgers equation and other problems with dissipative shocks. Here the effect of the dispersive term is retained at leading order and yields some interesting results.

For $i \geq 0$, we obtain linear ordinary differential equations for $U^{(i)}$

$$\mathscr{L}U^{(i)} = \mathscr{N}(U^{(i-1)}, \ldots, l^{(i)}, \ldots, K^{(i)}, \ldots), \qquad (43)$$

where

$$\mathscr{L} \equiv l^{(0)}\frac{\partial}{\partial\theta} + \frac{\partial}{\partial\theta}U^{(0)} + K^{(0)^2}\frac{\partial^3}{\partial\theta^3}.$$

Note that the inhomogeneous terms involve only known lower-order solutions.

To determine the dependence of $U^{(0)}$ on x and t, we use compatibility conditions for the existence of higher-order solutions to the inhomogeneous equation (43). For $i \geq 1$, these require that the inhomogeneous term $\mathscr{N}$ be orthogonal to the null space of the adjoint operator $\mathscr{L}^*$ defined by

$$-\mathscr{L}^* \equiv l^{(0)}\frac{\partial}{\partial\theta} + U^{(0)}\frac{\partial}{\partial\theta} + K^{(0)^2}\frac{\partial^3}{\partial\theta^3}. \qquad (44)$$

The only periodic solutions in the null space are 1 and $U^{(0)}$, and these lead to two equations averaged over θ and involving only x and t

differentiations. The third linearly independent solution is not periodic and is not used to derive an averaged equation.

3b. Leading-order Problem

It is possible to solve (42) explicitly for $U^{(0)}$ as a function of θ (hereafter the superscript must be dropped, since we deal only with the leading-order problem). We can integrate (42) twice to obtain

$$\tfrac{1}{2}lU^2 + \tfrac{1}{6}U^3 + \tfrac{1}{2}K^2U_\theta^2 = mU + n, \tag{45}$$

where m and n are integration "constants" depending on x and t. Requiring U to be periodic in θ imposes certain restrictions on K, l, m, and n, but these will not be stated here. After some manipulations we obtain the result

$$U(\theta, x, t) = -l - (l^2 + 2m)\{b - (b-a)cn^2[2\mathscr{K}(k)(\theta - \theta_0)\,|\,k]\}, \tag{46}$$

where cn is a Jacobi elliptic function, $\mathscr{K}$ is the complete elliptic integral of the first kind,

$$k^2 \equiv \frac{b-a}{c-a}, \tag{47}$$

and a, b, and c essentially correspond to the roots of (45) when $U_\theta = 0$ (see Chapter IV, Subsection 3c). All quantities on the righthand side of (46) are functions of l, m, and n, which in turn depend only on x and t.

To determine the dependence of U on x and t, it is convenient to introduce the averaged variables

$$P \equiv \langle U \rangle, \quad Q \equiv \langle (U - P)^2 \rangle, \quad R \equiv -\tfrac{1}{6}\langle (U - P)^3 \rangle - (P + l)Q, \tag{48}$$

where $\langle F \rangle \equiv \int_0^1 F(\theta)\,d\theta$. The two averaged compatibility equations do not form a closed system; it is necessary to introduce a third averaged equation to accomplish this. The third averaged equation is obtained by averaging the third conservation law—that is, the one with a conserved density that contains u^3. We obtain the following determined system of (complicated) first-order, quasilinear, partial differential equations in the averaged variables P, Q, and R:

$$P_t + PP_x + \tfrac{1}{2}Q_x = 0, \tag{49}$$

$$Q_t + 2QP_x + (P + 2\lambda + 2Q\lambda_Q)Q_x + (5 + 2Q\lambda_R)R_x = 0, \tag{50}$$

and

$$R_t + (2Q\lambda + 6R)P_x + [-3\lambda^2 - 6(Q\lambda + R)\lambda_Q]Q_x$$
$$+ [P - 6\lambda - 6(Q\lambda + R)\lambda_R]R_x = 0, \qquad (51)$$

where

$$\lambda \equiv P + l, \qquad (52)$$

$$\lambda_Q = \frac{\lambda^2(3Q\lambda + 5R) + \frac{1}{2}QR}{(2Q\lambda + 3R)(3Q\lambda + 5R)}, \qquad (53)$$

$$\lambda_R = \frac{3\lambda(3Q\lambda + 5R) - Q^2}{3(2Q\lambda + 3R)(3Q\lambda + 5R)}. \qquad (54)$$

This system of equations is invariant under change of sign of both x and t and under galilean transformation. Furthermore, after much algebra, one can show that this system of equations is hyperbolic. The condition leading to hyperbolicity turns out to be precisely that for the existence of three real roots of (45) when $U_\theta = 0$. This is undoubtedly not a mere coincidence; however, the connection between them is not clear.

3c. The Possibility of Shocks

One of the questions we have asked about this system of equations is whether or not shocks exist. A calculation of the jumps in the quantities P, Q, and R across a shock results in the conclusion that the jumps are of zero magnitude. This seemingly negative result does not exclude the possibility of a transition between two different states each characterized by given values for the averaged quantities. As indicated in Figure VIII.4, two uniform states with averages $P_1 = \langle u \rangle_1$ and $P_2 = \langle u \rangle_2$ may be joined by an oscillating transition region. To the left of this region the join is accomplished through small-amplitude oscillations that are essentially linear sinusoidal waves with average of $P = P_1$. To the right the join is through large oscillations. It has been observed that solitons always emerge from the right in numerical solutions; one can therefore interpret the oscillation bounding the transition region on the right as a soliton in the limit $\delta \to 0$. Averaging over a period of this bounding oscillation leads to the background value P_2 because the soliton has an infinite period.

Because the averaged quantities are continuous at the joins of the

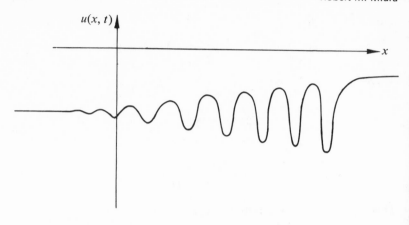

VIII.4.

transition region with the uniform states, the "shocks", interpreted now as the location where the structure of the solution changes abruptly, occur in pairs and are of zero strength. If, on the other hand, the entire oscillatory transition region is identified as a shock structure, then we must conclude from numerical solutions that the shock thickness will always increase indefinitely no matter how small $\delta \neq 0$. The shock, therefore, cannot be confined to a small region as in the dissipative systems, confirming again the conclusions of Chapter IV, Section 3. Similar results are reported by Yajima et al. (1966) for a different non-linear dispersive equation.

CHAPTER IX

Structure of Collisionless Shocks

Richard E. Meyer

A shock is essentially a wave of transition from one equilibrium state of a medium to a different equilibrium state. The first and simplest question is which pairs of equilibria can be connected by such a wave, and the answer is well known to be given by a set of shock relations. They express the conservation of mass and momentum and of further quantities that depend on the dynamical nature of the medium.

This type of analysis, pioneered by Rankine, Hugoniot, Rayleigh, and others, has been so well confirmed that it is rarely mentioned that conservation follows from an *assumption* about the transition, usually, that it is *steady* (in the frame of an appropriate observer). But how can we be assured that the physical medium possesses a steady transition mechanism? For ordinary shocks in gases, such a mechanism is now very well established; it depends essentially on the mechanism of molecular collisions. The answer is less obvious for plasma shocks like the bow shock of the earth's magnetosphere in the solar wind (Figure IX.1). The observational evidence on such transitions is still very limited. If they are steady, the mechanism cannot depend on collisions, because the number density of the particles in the solar wind is too small. If they are unsteady, on the other hand, then the whole basis for shock relations collapses, for there is then no a priori reason why even mass or momentum should be conserved.

(In case this is felt to involve resort to a very outlandish physical model to make the point, it may be noted that the studies to be reported

This work was supported in part by a National Science Foundation grant.

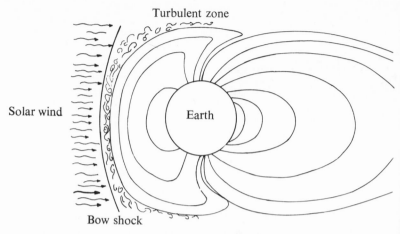

Turbulent zone

Solar wind

Earth

Bow shock

IX.1. Sketch of the bow shock of the earth's magnetosphere in the solar wind.

here were originally concerned with the commonplace undular jumps and bores of hydraulics (see Chapter IV). Gardner and Morikawa (1960) noted an analogy between them and plasma shocks, to which the studies turned when the hydraulic problem (Meyer, 1967a) was found to be even more difficult and less transparent.)

1. Steady Shocks

For collisionless plasma shocks, doubts about steadiness were crystallized by Morton's paradox on plane shocks. In a plane wave, physical quantities depend on only one space coordinate. If it is also steady, such a wave is therefore governed by ordinary differential equations; and their solution set can be completely discussed, even for a quite complicated plasma model, by the methods of the qualitative theory of ordinary differential equations. This was done by Morton (1962) for a reasonably realistic, two-fluid model of collisionless plasma. For a shock front normal to the propagation direction and transverse to the magnetic field, the solution set can be classified in the plane of two parameters: the ratio β of fluid pressure p_f to magnetic pressure p_M and the shock strength ϵ. It is found that there is a critical curve $\epsilon = \epsilon_c(\beta)$ in this plane such that the model possesses semi-discontinuous,

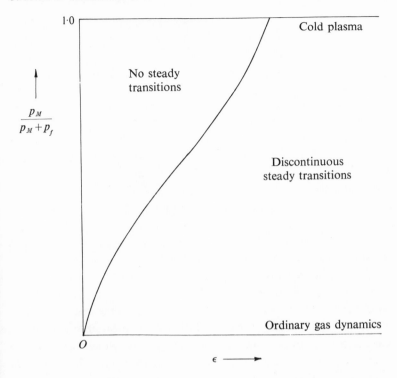

IX.2. Domains of existence of solutions to the steady state equations for a collisionless plasma.

steady transition solutions for $\epsilon > \epsilon_c$; but for smaller shock strengths, $\epsilon < \epsilon_c$, only periodic or solitary wave solutions exist (see Figure IX.2). The latter cannot represent a transition from one equilibrium state to a different one, and hence, for sufficiently *small* shock strength ϵ, no plane, steady shocks exist. Now, it is plausible that any given model becomes inadequate as the shock strength becomes too large, but the nonexistence of *weak* shocks is a quite unusual difficulty!

The only exception is the limit $p_M/p_f \to 0$, which corresponds to ordinary gas dynamics and admits steady shocks of arbitrary strength. The paradox is most pronounced in the limit $p_f/p_M \to 0$, which corresponds to cold plasma, and a somewhat simpler plasma model is therefore likely to suffice for its study. Before turning to this, however, it will be helpful to discuss a more general aspect of the difficulty indicated by Morton's paradox.

2. Sets of Limits

Once it is realized that the existence of steady, collisionless shocks is not certain, especially when the shocks are weak, it follows that the questions under study involve a double limit in which the shock strength $\epsilon \to 0$ and the time $t \to \infty$. This is because we would not expect to observe the shock readily until it settles down to some degree of near-steadiness, at least, in the frame of some galilean observer.

This double limit is, of course, very familiar, since much of mathematical physics involves both the notions of small amplitude and large time. A standard procedure is therefore very well established; it is to let $\epsilon \to 0$ first, to linearize the problem, and then to work out the solution asymptotically for large time.

It is not remarked very often that experimental physics is usually concerned with the limits in the opposite order. In the case of a shock, for instance, the observations concern almost invariably a shock of fixed strength, however small, over times after it has settled down to a sufficient degree. Much of theoretical physics is thus based on the implicit trust that the two limits commute. It is perhaps neither surprising that this principle should have served very well over quite a number of generations, nor that a point has finally been reached where the problems to be faced are so sophisticated that exceptions to the "rule" are encountered.

The result of applying the classical $\lim_{t \to \infty} \lim_{\epsilon \to 0}$ will appear automatically below in the search for the correct limit. That result was obtained independently by several authors, by classical or by quite modern asymptotic methods, but not then published, because it is not satisfactory. Much worse, that result looks eminently correct and plausible, at first sight, but is wrong nonetheless. This was a first indication that Morton's paradox might be subtle and might resist the excellent methods expounded in the other chapters of this book.

Since the hard order of the double limit,

$$\lim_{\epsilon \to 0} \lim_{t \to \infty}$$

has to be faced, there is no hope of linearization, and there does not appear to be any established method for nonlinear double asymptotics. The approach (Meyer, 1967b, c) to be presented here was motivated by the following idea. The required double limit can be represented symbolically in the plane of the variables $1/t$ and ϵ by Figure IX.3. The limit $t \to 0$ is to be considered for fixed ϵ, and this is to be done for a sequence of values of ϵ decreasing towards zero. This double limit should be interpretable as a *limit of single limits* represented in the t^{-1}, ϵ-plane by sequences of curves approaching the origin more and more steeply as shown in Figure IX.4. Any one such curve would represent a limit in which $1/t \to 0$ in some relation to ϵ—in other words, in which the time

$$t = t_\theta \theta(\epsilon),$$

with fixed t_θ and a suitable function $\theta(\epsilon)$ tending to ∞ as $\epsilon \to 0$.

A first impulse might be to consider for $\theta(\epsilon)$ the family of all powers ϵ^k with real k. But of course, logarithms and exponentials (and combinations of all these) need also be admitted, and this family is quite complicated to handle and yet not satisfactorily rounded out. Simpler and better progress can be made in the hunt for limit points of sets of limits by the help of Kaplun's (1967) investigations into the foundations underlying inner-outer expansions and many other asymptotic methods.

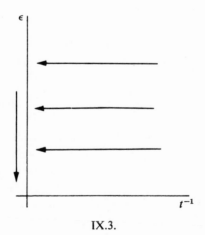

IX.3.

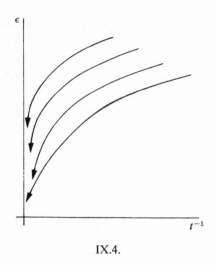

IX.4.

Let Ω denote the set

$$\Omega = \{f(\epsilon) \in C(0, 1) \mid f(\epsilon) > 0 \text{ for } 0 < \epsilon \le 1\}$$

of all positive, continuous functions of ϵ defined for $0 < \epsilon \le 1$ (i.e., of all possible parameters, small or large). For any function $\gamma(\epsilon)$ in Ω an *order class* can be defined by the equivalence relation

$$\text{ord } \gamma = \{\mu(\epsilon) \in \Omega \mid \lim_{\epsilon \downarrow 0} \frac{\mu(\epsilon)}{\gamma(\epsilon)} \,\exists \text{ and} \ne 0\}$$

and functions in the same class need not be distinguished for a study of the limit $\epsilon \to 0$. Among these classes, an ordering can be introduced by the definition

$$\text{ord } \delta < \text{ord } \gamma \quad \text{means} \quad \lim_{\epsilon \downarrow 0} \frac{\delta(\epsilon)}{\gamma(\epsilon)} = 0.$$

But this yields only a partial ordering, since, for instance, $2 + \sin(1/\epsilon)$ $\in \Omega$, but its order class neither $=$ nor $>$ nor $<$ ord 1.

The set of all order classes is thus seen to be much larger than the set of real numbers. A topology can be introduced on it by the following two definitions. A set S of order classes is *convex* means that ord $\gamma \in S$, ord $\delta \in S$, and ord $\gamma <$ ord $\mu <$ ord δ imply that also ord $\mu \in S$. In other words, if a convex set contains two mutually orderable classes, then it also contains every class that can be ordered between the two; the set has no gaps. A set S of order classes is *convex-open* means it is convex and if ord $\mu \in S$, then $\gamma(\epsilon)$ and $\delta(\epsilon)$ in Ω exist such that ord $\mu <$ ord $\gamma \in S$ and ord $\mu >$ ord $\delta \in S$. In other words, a convex-open set contains, with every member ord μ, also members to the right and to the left of ord μ (Kaplun and Lagerstrom, 1957). A set is *open* means that it is a union of convex-open sets. For the following, it will be useful to add the notion of the *right set*

$$R_\gamma = \{\text{ord } \mu \mid \text{ord } \gamma < \text{ord } \mu\}$$

of $\gamma \in \Omega$—that is, the set of all order classes to the right of ord γ (Meyer, 1967b). [This set corresponds, in the more naive sense of Figure IX.4, to the family of all curves plunging to the origin more steeply than the given curve $t = t_\gamma \gamma(\epsilon)$.]

Now consider any function $F(x; \epsilon)$ of a variable x and a positive parameter ϵ. The most common concern is with a vector variable x of which the size $|x|$ is measured in terms of the usual euclidean distance, and that will also be the case needed below. F may be any function the size of which is defined by some norm; a particular, real, vector-valued function and norm of interest will emerge in Section 6 from the formulation of the plasma problem.

The main concern of modern asymptotics is with nonuniformity of approximation arising when a single limit process is insufficient to cover the closure of the x domain. For the moment, assume $x = 0$ to be the only exceptional point, so that $F(x; \epsilon)$ is defined, say, for all $|x| > 0$ and for $0 < \epsilon \le 1$. To cover a neighborhood of $x = 0$, it may then be necessary to consider the dependent variable F as a function, not only of x and ϵ, but also of a different variable y and ϵ, where $y = y(x, \epsilon)$ already depends on ϵ. To avoid prohibitive generality, assume that F depends, near $x = 0$, primarily on $|x|$, so that a new variable $x_\delta = x/\delta(\epsilon)$ suffices, where $\delta \in \Omega$ is a parameter dependent only on ϵ. More precisely, this involves a transformation of variables from x, ϵ, F to x_δ, ϵ, G defined by

$$x_\delta = x/\delta(\epsilon), \qquad G(x_\delta; \epsilon) = F(x; \epsilon)$$

so that G has the same function values as F at corresponding points.

This is a "stretching transformation," and Kaplun's (1967) idea was to consider simultaneously the set of such transformations for all $\delta \in \Omega$ and a set of associated limits. Following Freund (1972), we adopt the following definition: *if, for all closed intervals* $[a, b] \subset (0, \infty)$, *the function* $G(x_\delta; +0) \equiv \lim_{\epsilon \downarrow 0} G(x_\delta; \epsilon)$ *exists and is approached uniformly on* $\{x_\delta \mid \lvert x_\delta \rvert \in [a, b]\}$, *then* $\lim_\delta F(x; \epsilon) \equiv G(x_\delta; +0)$.

In other words $\lim_\delta F$ is $\lim F(z\delta(\epsilon); \epsilon)$ as $\epsilon \to 0$ with, not x, but z fixed. For instance, if $\delta(\epsilon) \to 0$ as $\epsilon \to 0$, then also $x \to 0$ in the limit process $\lim_\delta$ in such a manner that x/δ tends to a nonzero limit vector. A similar definition with more restrictive uniformity requirement was considered by Meyer (1967b); Kaplun (1967) did not make quite clear which definition he had in mind.

The "speed" with which $\lvert x \rvert \to 0$ as $\epsilon \to 0$ in the $\lim_\delta$ process is seen to depend only on ord δ. In fact, the whole definition of Kaplun limit depends only on ord δ, since (Freund, 1972) if $\lim_\delta F = 0$ and ord $\gamma =$ ord δ, then $\lim_\gamma F = 0$.

3. Double Asymptotics

In mathematical physics, a parameter usually enters into the formulation of a problem only in a few places. For each problem, the Kaplun limits therefore fall naturally into families within which the distinction between different limits is less substantial. The following definition (Kaplun and Lagerstrom, 1957) is then helpful: $H(x; \epsilon)$ *is a* $\Delta(\epsilon)$-*approximation to* $F(x; \epsilon)$ *on a convex set* S *means* $\lim_\delta \lVert \Delta^{-1}(F - H) \rVert = 0$ *for all* ord $\delta \in S$.

These definitions support the *Kaplun Extension Theorem* concerning the foundation of matching of asymptotic expansions (Kaplun and Lagerstrom, 1957; Freund, 1972). Their application to large-time asymptotics still requires the removal of a notational awkwardness, since the functions envisaged are not really expected to tend to proper limits as $t \to \infty$, but rather to approach simpler functions of t. On the other hand, the standard notation $\phi(t) \sim \Psi(t)$ does not lend itself to usage for double limits. As a compromise, the notation $\Psi(t) = \lim^*_{t \to \infty} \phi(t)$ will be used to express that $\Psi(t)$ is an asymptotic approximation to $\phi(t)$—in other words,

$$\lVert \Psi - \phi \rVert \to 0 \qquad \text{as } t \to \infty.$$

(With suitable choice of Ψ' and ϕ, this covers approximation to any order.)

Now consider a function $f(t, \epsilon)$ of two real variables, for simplicity, and the "double limit"

$$\lim_{\epsilon \downarrow 0} \lim_{t \to \infty} * f(t, \epsilon).$$

We may then ask for an asymptotic approximation, as $\epsilon \to 0$, to an asymptotic approximation

$$\lim_{t \to \infty} * f(t, \epsilon) = g(t, \epsilon)$$

to f as $t \to \infty$. Precisely stated, an asymptotic $\Delta(\epsilon)$-approximation to $\lim^*_{t \to \infty} f(t, \epsilon)$ as $\epsilon \to 0$ means a function $h(t, \epsilon)$ with the property that, given any $\eta > 0$, there exist positive numbers A_η and T_η such that $0 < \epsilon < A_\eta$ and $t > T_\eta$ imply

$$||\Delta^{-1}[h(t, \epsilon) - \lim_{t \to \infty} * f(t, \epsilon)]|| < \eta.$$

For a double limit, this is not the only type of approximation that can be envisaged. While the definition implies that the difference between $\lim^* f$ and f decreases progressively as time grows, for any fixed ϵ, the same is not here demanded of the difference between h and f. But a uniform bound (decreasing progressively with ϵ) imposed on that difference suffices for many physical applications. In fact, for the purpose of characterizing the ultimate, as opposed to any transient, behavior in time for arbitrarily small fixed ϵ, the following concept due to Freund (1972) suffices: $h(t, \epsilon)$ is an *eventual* $\Delta(\epsilon)$-approximation to $\lim^*_{t \to \infty} f(t, \epsilon)$ as $\epsilon \to 0$ if, given any $\eta > 0$, a number $A_\eta > 0$ and positive function $T_\eta(\epsilon)$ on $(0, A_\eta]$ exist such that $0 < \epsilon < A_\eta$ and $t > T_\eta(\epsilon)$ imply

$$||\Delta^{-1}[h(t, \epsilon) - \lim_{t \to \infty} * f(t, \epsilon)]|| < \eta.$$

This concept, moreover, may often be used to relate a double limit to a single Kaplun limit by the following theorem of Freund (1972):

THEOREM. *$h(t, \epsilon)$ is a $\Delta(\epsilon)$-approximation to $f(t, \epsilon)$ on a right set if, and only if, it is an eventual Δ-approximation to $\lim^*_{t \to \infty} f$ as $\epsilon \to 0$.*

Here right set is interpreted in terms of $\lim_\theta$ and $t_\theta = t/\theta$ in the place of $\lim_\delta$ and $x_\delta = x/\delta$ in the definitions given earlier. The more restrictive limit definition of Meyer (1967) supports the same theorem with "asymptotic" in the place of "eventual." but it also implies that then $h(t, 0) \equiv f(t, 0)$ for $t > T_0$ (Freund, 1970) so that the range of practical application of that theorem must be too narrow.

Some progress has thus been made in the exploration of the nonlinear double asymptotics that is bound to play an increasing role in mathematical physics. The phrase "double limit" is seen to be merely a doorway to many distinct approximation concepts. Among them, an eventual approximation can be distinguished from transient approximations by the simple criterion of Freund's theorem. In practice, it thus suffices to analyze single Kaplun limits. If the result is seen to approximate on a right set, an eventual approximation is assured. If not, then the result cannot be more than a *transient* asymptotic approximation.

The discussion of double asymptotics has here been phrased in terms of just two scalar variables, for simplicity, but there is no difficulty in transcribing it below to the case where further variables are present. It will also be more convenient to use the function $\tau(\epsilon) = 1/\theta(\epsilon)$, and the criterion will then be given by Freund's theorem with $\lim_\theta$ and right set replaced by $\lim_\tau$ and left set

$$L_\gamma = \{\text{ord } \mu \mid \text{ord } \mu < \text{ord } \gamma\}.$$

4. Plasma Model*

It will be helpful to use the simplest collisionless plasma model exhibiting Morton's paradox, and Figure IX.2 indicates that this is a plasma of electrons and singly charged ions that is quite *cold*, so that there is no random motion. The ion and electron velocities are then ordinary vector fields $\mathbf{u}_+(\mathbf{x}, t)$ and $\mathbf{u}_-(\mathbf{x}, t)$, respectively, and number conservation is expressed by

$$\frac{D_+ n_+}{Dt} + n_+ \text{ div } \mathbf{u}_+ = 0, \qquad \frac{D_- n_-}{Dt} + n_- \text{ div } \mathbf{u}_- = 0,$$

where n_+, n_- are the respective number densities and

* The reader who is not interested in the physics of collisionless shocks may wish to proceed directly to the model equations on page 247.

$$\frac{D_{\pm}}{Dt} = \frac{\partial}{\partial t} + \mathbf{u}_{\pm} \cdot \text{grad}$$

are the convective rates of change observed when one follows the "fluids" in their respective motions. The completely cold plasma has no fluid pressure; conservation of momentum is then expressed by

$$\frac{m_{+}D_{+}\mathbf{u}_{+}}{Dt} = e(\mathbf{E} + \mathbf{u}_{+} \times \mathbf{B}), \qquad \frac{m_{-}D_{-}\mathbf{u}_{-}}{Dt} = -e(\mathbf{E} + \mathbf{u}_{-} \times \mathbf{B}),$$

where $m_{\pm}$ and $\pm e$ denote the respective masses and charges.

The plasma will also be assumed *charge-neutral* (because it is non-relativistic), so that $n_{+} = n_{-}$. Furthermore, attention will be restricted to *plane waves*, for which the physical quantities depend on time and only one space coordinate x. Then, if the number density is not trivially uniform in space, any difference between the x components of the ion and electron velocities must lead to a charge separation, which is ruled out by the charge neutrality (because such a difference $u_{1+} - u_{1-}$ would, like $n_{+} - n_{-}$, decay on a relativistic time scale). Accordingly,

$$\frac{D_{+}}{Dt} = \frac{D_{-}}{Dt} = \frac{D}{Dt} = \frac{\partial}{\partial t} + u_{1}\frac{\partial}{\partial x},$$

where u_{1} is the common x component of the ion and electron velocities. Any velocity difference

$$\mathbf{u}_{+} - \mathbf{u}_{-} = \mathbf{v}$$

must therefore be perpendicular to the x direction; it implies a current

$$\mathbf{J} = ne\mathbf{v}.$$

It is convenient to rewrite conservation of momentum in terms of $\mathbf{v}$ and of the mass-average velocity

$$\mathbf{u} = \frac{m_{+}\mathbf{u}_{+} + m_{-}\mathbf{u}_{-}}{m_{+} + m_{-}},$$

which yields

$$(m_{+} + m_{-})\frac{D\mathbf{u}}{Dt} = e\mathbf{v} \times \mathbf{B},$$

$$m_{+}m_{-}\frac{D\mathbf{v}}{Dt} = e(m_{+} + m_{-})(\mathbf{E} + \mathbf{u} \times \mathbf{B}) - e(m_{+} - m_{-})\mathbf{v} \times \mathbf{B}.$$

Then the first component of the equation for $\mathbf{v}$ defines the "charge-separation field" E_1, since charge neutrality has been seen to imply $v_1 \equiv 0$. For plane waves, moreover, Maxwell's equation div $\mathbf{B} = 0$ shows B_1 to be independent of x.

Our concern is with solutions of these and Maxwell's equations, which describe transitions from one equilibrium state of the plasma to a different one. Without loss of generality, therefore, there is one inertial frame, the rest frame, in which the plasma is at rest (and thus $\mathbf{v} = 0$) at $x = +\infty$. It will be enough here to consider the simplest case, that of a normal, transverse shock; the results for the general case are given by Olson (1972). That a shock is normal means that it propagates in the direction normal to the planes of constant phase. Our concern is therefore only with solutions that approach steadiness in the frame of an observer who travels with respect to the rest frame with a fixed velocity U in the direction of increasing x. In the frame of this observer, the equilibrium state into which the shock propagates is characterized by

$$u_1 \to -U, \qquad u_2{}^2 + u_3{}^2 \to 0 \qquad \text{as } x \to +\infty.$$

Here $U > 0$, without loss of generality, because a solution is sought in which plasma, by passage through the shock, is brought from one state to another.

That a shock is transverse means that the magnetic field B_∞ in the equilibrium state ahead of the shock is perpendicular to the propagation direction of the shock. For equilibrium, it must also be perpendicular to the electric field there. The direction of that magnetic field will be chosen as that of increasing z; then B_1, B_2, and E_3 each approach 0 as $x \to +\infty$. It follows that $B_1 \equiv 0$, since it is independent of x. Symmetry makes it plausible, moreover, that the momentum equations and Maxwell's equations

$$\text{curl } \mathbf{E} = -\mathbf{B}_t, \qquad \text{curl } \mathbf{B} = \mu \mathbf{J},$$

can then be solved with $u_2 \equiv u_3 \equiv B_2 \equiv E_3 \equiv v_3 \equiv 0$, and it has been shown (Olson, 1972) that any solution must have this property in the limit of zero amplitude.

We are thus left with only one average-velocity component u (in the x direction), only one magnetic field component B (in the z direction), and only one difference-velocity component v (in the y direction). On assuming that the difference velocity is twice continuously differentiable, it follows from curl $\mathbf{E} = -\mathbf{B}_t$ and the difference-momentum equation that $v_x + e(m_+ + m_-)B/(m_+ m_-)$ satisfies the same conservation law as n. On assuming uniqueness, that implies the first integral (Gardner and Morikawa, 1960)

$$v_x = e\,\frac{m_+ + m_-}{m_+ m_-}\left(\frac{B_\infty}{n_\infty}\,n - B\right), \tag{1}$$

which, together with

$$n_t + (nu)_x = 0, \tag{2}$$

$$u_t + uu_x = \frac{evB}{(m_+ + m_-)}, \tag{3}$$

$$B_x + \mu env = 0, \tag{4}$$

forms the system of governing equations. The boundary conditions ahead of the shock are

$$u \to -U, \quad v \to 0, \quad n \to n_\infty, \quad B \to B_\infty \quad \text{as } x \to +\infty \tag{5}$$

for all t, with given constants n_∞, B_∞ and constant $U>0$ to be determined so that a solution is possible that approaches steadiness arbitrarily closely and tends to an equilibrium state with $B \neq B_\infty$ as $x \to -\infty$.

5. Stretching

The equations (1) to (4) are a nonlinear system of two conservation laws and two additional equations, which is quite typical of systems encountered in mathematical physics. Any method for their study in the nonlinear limit of near-steadiness at fixed small amplitude should therefore prove helpful also in many other fields. The method to be formulated now relies on the result (Section 3) that sufficient information on the double limit can be obtained from sets of single Kaplun limits in which amplitude and rate of time dependence of the solution tend to zero in some relation.

Most such limits must be anticipated to involve some degree of singularity or degeneration of the solution, so that stretching transformations are necessary for their study. Accordingly, we introduce the quite general transformation

$$x' = \frac{\delta x}{l}, \qquad t' = \frac{\gamma \delta t U}{l},$$

$$u(x, t) = -U + \epsilon U u'(x', t'; \epsilon),$$
$$v(x, t) = \theta U v'(x', t'; \epsilon),$$
$$n(x, t) = n_\infty + \kappa n_\infty n'(x', t'; \epsilon), \qquad (6)$$
$$B(x, t) = B_\infty + \beta B_\infty b'(x', t'; \epsilon),$$

$$l^{-1} = e\left(\mu n_\infty \frac{m_+ + m_-}{m_+ m_-}\right)^{1/2},$$

to nondimensional primed variables, where δ, γ, θ, κ, and β are non-dimensional positive functions of ϵ only.

Since the new dependent variables are defined as perturbations from the equilibrium ahead of the shock, the restriction to small amplitude implies that ϵ, θ, κ, and β are all small parameters. But we abstain from the usual procedure of assuming that they are all equal or are certain plausible powers of ϵ or are otherwise related in some definite, pre-conceived way.

Restrictions arise, even with so general a stretching transformation, primarily in regard to the independent variables. There is no a priori reason why a single length scale l/δ should suffice to characterize the whole shock transition. For a classical gas-dynamical shock a single scale does suffice, but for the plasma shock it will emerge not to be always enough, and dispersion generally tends to lead to multi-scale problems, as seen in most of the other chapters. It is then advisable to attempt to understand the scales one after another, and in the instance of weak plasma shocks, the difficulties associated with one scale are ample for the present account. Attention will therefore be restricted here to the *head of the wave*, defined as its very front and as much behind the very front as is characterized by a single length scale.

It follows that a tail boundary condition specifying the plasma equilibrium in which the transition ends, cannot generally be imposed in the present analysis. The resulting degree of nonuniqueness, however, will not turn out to be prohibitive.

On the other hand, an overall property of the length scale $l/\delta = L$ for the head of the wave can be predicted a priori. A linearized analysis c

(1) to (4) shows (Olson 1972) their dispersion relation to be such that the longest waves travel fastest, and those waves will therefore come to form the front of the shock transition, as it gradually approaches near-steadiness with increasing time. It is thus safe to anticipate that the length scale of the head of the wave will grow to be long compared with the physical dimension l, so that δ is also a small parameter, and we shall profit from this assumption to reduce the volume of work.

A single time scale cannot be expected either to cover the development of the shock transition, but our present interest centers upon the description of the head of the wave during the ultimate, near-steady stage of its development. The time scale $T = l/(U\gamma\delta)$ in (6) must therefore be taken to be the smallest relevant time scale during that stage. Furthermore, the natural interpretation of normal unsteadiness, neither large nor small, in physics is that the relevant time scale is the ratio L/U of the relevant length and velocity scales. Near-steadiness therefore means that $T \gg L/U$—in other words, that γ is also a small parameter.

To sum up, the considerations of this section have shown that

$$\beta, \gamma, \delta, \theta, \kappa \in \Omega_0 = \{\rho \in \Omega \mid \rho \to 0 \text{ as } \epsilon \to 0\}$$
$$= \{\rho \in C(0, 1] \mid \rho > 0 \text{ and } \rho \to 0 \text{ as } \epsilon \to 0\} \quad (7)$$

by the definition of the problem, except in the case of δ, where this fact is partly an assumption based on the dispersion relation.

The restriction to a single time scale precludes generally the formulation of a standard initial-value problem. The resulting degree of nonuniqueness is also not likely to be prohibitive. In fact, experimental initial conditions for the development of a collisionless plasma shock are not usually controllable in detail and may well be effectively unobservable. (Experimental scientists, moreover, will tend to support similar remarks in regard to many processes of physics.) To approach the same question from another direction, applied mathematics tends to accept as the most natural initial condition for the development of a transition wave of (1) to (4) a simple step-function distribution of u, v, n, and B. But of course, that is the opposite of a physically natural distribution. Rather, it represents a canonical initial condition based on the implicit premise that the solution for it is likely to be representative of the physically observed solutions (grown from whatever actual initial distributions occurred, which probably differed substantially from step functions). In short, the use of a canonical (rather than measured) initial condition implies, if not a knowledge, then an assumption that the late-time features of primary interest are independent of that initial condition. It is at least as honest, then, to inquire what solutions could emerge eventually, without consideration of initial conditions; and that approach will be chosen here.

Time thus enters into the problem only through the differential operators of the conservation laws (2), (3), and the only meaning assignable to the notion of a large time scale $T = l/(U\gamma\delta)$ is that the small parameter

$$\tau(\epsilon) = \gamma\delta$$

characterizes the nondimensional time derivatives at the head of the wave at the times under consideration. Direct, large-time asymptotics is thus replaced by asymptotics for a close approach to steadiness. In some instances, the results below are detailed enough to permit illuminating deductions in regard to actual time asymptotics. But the primary physical concern of the analysis is with near-steadiness, in any case (and the references to time asymptotics in Section 2 use the standard terminology only for convention's sake).

The abandonment of an explicit initial condition also increases the flexibility of the stretching (6), since the origin of t' can be chosen as late as may be convenient, and in any case, during a stage of development of the shock when $T = l/(U\gamma\delta)$ is already the smallest relevant time scale at the head of the wave. The analysis will, moreover, cover large time intervals—namely, at least, any finite interval of $t' = t/T$.

The propagation velocity U, finally, must be dimensionally related to the Alfvén speed (see Chapter I, Subsection 4d).

$$V_A = B_\infty [\mu n_\infty (m_+ + m_-)]^{-1/2}.$$

In fact, U/V_A must be positive by definition (Section 4) and bounded, since the plasma is assumed nonrelativistic, and if U did not tend to a limit as $\epsilon \to 0$, the small-amplitude limit would be considered physically inadmissible. We may thus write

$$U = V_A(1 - \lambda)^{-1/2}, \qquad 1 > \lambda(\epsilon) \in C[0, 1]. \tag{8}$$

6. Scale Conditions

This lengthy interpretation of the stretching transformation (6) still does not go quite far enough in clarifying the stretching for the purposes at hand. Before we consider this, however, it is more urgent to discuss the direction and aim of these considerations. That may be done by reference to the familiar method of inner-and-outer expansions or "stretching-and-matching," in which a particular stretching transformation is proposed and then justified by matching. But in fact, such justification is rarely possible in nontrivial problems, because it would require matching to all orders, to the extent of an existence proof. In problems of nonlinear, partial differential equations the complexities tend to prohibit matching even to the third order; hence, the program of justifying a proposed stretching by matching is usually an empty one.

Some interest, therefore, attaches to methods by which a stretching could be determined independently of matching. Such methods might be applicable also to multi-scale asymptotics, which employs stretching without direct matching. They would be of particular interest in cases, like the present, where the stretching indicated by plausibility and experience leads to results that look very plausible but are incorrect.

The particular aim of this section is to formulate a method by which the *stretching* transformation for a problem can be *determined uniquely, independently of matching.* Such a uniqueness proof may avoid some of the formidable obstacles often impeding proof of existence and asymtotic property and may then constitute a useful asymptotic method complementary to "stretching-and-matching" and to multi-scale approximation. The scanning of large sets of limit processes implicit in such a uniqueness proof also harmonizes with the purpose of identifying an eventual approximation in double asymptotics (Section 3).

These aims require a flexible stretching transformation, representing a net cast very wide, and this makes careful scrutiny of the generality and limitations of the transformation (6) relevant. The next question, however, concerns the principles that may permit the unique determination of the stretching transformation (6). It will be proposed now that these need amount to little more than the expression, with some precision, of what is meant by "stretching" or "ordering" or physical "scale."

The basic purpose of stretching is to provide a magnifying glass (or reverse of it) fashioned precisely so as to resolve the degeneracy of a limit of the solution of a physical problem. In the case of the plasma shock, for instance, there is no reason to expect that the dimensional magnetic field $B(x, t)$ has a limit as the amplitude $\epsilon \to 0$. But it is a widely accepted premise of physics that—unless the contrary be demonstrated rigorously—units (possibly dependent on ϵ) can be found such that the magnetic field perturbation $b'(x', t'; \epsilon)$ measured in those units will have a limit.

In this connection, it is understood that "limit" includes more than boundedness of $b'(x', t'; \epsilon)$. A physical quantity such as a magnetic field may degenerate, even at unit amplitude, in that its frequency grows beyond any bound. That is why the magnifying glass applies to the units of length and time as well as to any other physical units. In short, the idea is that, while the limit of the original field $B(x, t)$ may be quite degenerate, units can be found in terms of which $b'(x', t'; \epsilon)$ has as a limit an ordinary, reasonably smooth function.

One way in which this may be formulated for the plasma problem is to let

$$\Gamma^n = \{ f(x', t'; \epsilon) \in C^n(E) |$$

$$\lim_{\epsilon \to 0} \partial^k f(x', t'; \epsilon)/\partial x'^j \, \partial t'^{k-j} = \partial^k f(x', t'; 0)/\partial x'^j \, \partial t'^{k-j}$$

$$\in C(E) \text{ for } 0 \le j \le k \le n \},$$

where E denotes the set of all x', t' (much in constrast to the set of all x, t). In other words, these functions have a limit in $C^n(E)$, and the limits of their derivatives agree with the derivatives of the limit. Then the first of three Scale Conditions for the determination of the stretching (6) to be assumed here may be stated as

$$u', v', n', \text{ and } b' \quad \text{are in } \Gamma^2. \tag{I}$$

It is to be remarked that (I) clearly amounts to the choice of a norm in which the small-amplitude limit will be assumed to exist. This choice is far from the only possible one. However, stretching is a device to reduce the need for weak solution concepts, so it is natural to explore the physical subject, for a start, in terms of strict solutions. But such a decision to interpret the governing differential equations (1)–(4) more literally than symbolically, automatically determines the appropriate norm: since (1)–(4) are a first-order system, a strict solution is the same as a solution in Γ^1. Actually, an assumption on second derivatives enters into the derivation (Section 4) of (1), and Γ^1 was replaced by Γ^2 in Meyer (1967c) for the sake of a saving in labor without, so far, any indication of a significant loss of generality.

The more important point is that Γ^2 in (I) is not a particular function class corresponding to a particular norm in regard to the governing equations (1)–(4), but rather depends on the choice to be made from the large set (6) of stretching transformations. It is not assumed that the governing equations (1)–(4) have a strict solution in any limit of zero amplitude. The condition (I) postulates only that there is *some* zero-amplitude limit in which the degeneracy of the solution of (1)–(4) is decipherable by the help of *some* stretching transformation. That is the basic premise of the analysis discussed in this chapter.

The scale condition (I) may be called a regularity condition because it assures enough stretching to resolve any degeneracy of the limit, within the limitations of (6) already discussed. It must be complemented by a "nontriviality condition" guarding against excessive stretching. For instance, however singular $B(x, t)$ may become as $\epsilon \to 0$, it is always possible to choose the scale $\beta(\epsilon)B_\infty$ in (6) so large that $b' \to 0$ with ϵ for all x', t'. Such a choice is obviously not intended. A way to exclude it for the plasma problem is to require that

$$u',\ v',\ n',\ b' \quad \text{and} \quad (|u'_{t'}| + |v'_{t'}| + |n'_{t'}| + |b'_{t'}|)\ \text{are not trivial.} \qquad \text{(II)}$$

Here it is understood that a function $f(x', t'; \epsilon)$ is *trivial* means lub_V $|f| \to 0$ as $\epsilon \to 0$ on every open bounded set $V \subset E$.

That (II) includes a condition on derivatives is not unexpected, since overstretching of the independent variables would express itself in the loss of all variation of the dependent variables. It is less expected that no condition on the x derivatives is included, but that turns out to be already implied by the other conditions and the boundary condition (5). In the absence of explicit initial conditions, on the other hand, uniqueness of the stretching cannot be expected without a condition on the time derivatives. The cagey definition of trivial is designed to avoid the ambiguities arising from any further independent limits. For instance, the key boundary condition (5) is utterly ambiguous as it stands because it involves such a limit. Since x' is, by definition, the properly scaled space variable for the head of the wave, (6) permits a clearer formulation of (5) as

$$u', \ v', \ n', \text{ and } b' \to 0 \qquad \text{as } x' \to +\infty \quad \text{for all } t', \qquad (9)$$

uniformly in ϵ.

While the scale conditions (I) and (II) give a complete characterization of the idea of scale or stretching, they are not always sufficient to determine it, because stretching by itself usually involves loss of a boundary condition. When it does turn out possible to apply a tail boundary condition for the shock consistently on the length scale L, so that the head of the wave can cover the entire transition, that boundary condition will be taken as

$$b'(x', \ t'; \ 0) \to b_1 = const. \text{ as } x' \to -\infty \quad \text{for all } t'. \qquad (10)$$

When such a boundary condition cannot be applied, however, it is necessary for uniqueness to add to (I) and (II) a condition assuring that the head of the wave is a part of a larger transition process and is a substantial part of that process. The condition of Meyer (1967c) is

$$b'(x', \ t'; \ 0) \text{ remains bounded as } x' \to -\infty \quad \text{for all } t'. \qquad (III)$$

7. Transients

The immediate question is why these meager principles should suffice to determine the stretching? The claim is that (1)–(4), (9), and (I)–(III) determine the six unknown parameter functions $\beta(\epsilon)$, $\gamma(\epsilon)$, $\delta(\epsilon)$, $\theta(\epsilon)$, $\lambda(\epsilon)$, and $\kappa(\epsilon)$ uniquely. This turns out to be slightly exaggerated. The size of the set of limits to be scanned is that of the set of all order classes (Section 2) to the (cartesian) power 6, and a quite sufficient success will be to reduce the possible stretchings to a finite number so that their relative merits can be individually compared. To show that this claim is realistic requires proof in one thoroughly nontrivial case (and the plasma problem qualifies), but the volume of such an analysis necessarily exceeds the scope of the present account.

A compromise is to show the flavor of the analysis at the instance of just one of the governing equations. Substitution of (6) into the conservation law (2)—and omission of the primes, henceforth, since only the stretched variables will be considered in the following sections—gives

$$n_x - \left(\frac{\epsilon}{\kappa}\right) u_x = \gamma n_t + \epsilon(nu)_x, \tag{11}$$

and the righthand side here is trivial by (I) and because γ, $\kappa \in \Omega_0 = \{\rho(\epsilon) \in \Omega \mid \rho \to 0 \text{ as } \epsilon \to 0\}$.

Neither n_x nor u_x, on the other hand, can be trivial because it is not difficult to check (Meyer, 1967c; Olson, 1972) that a function in Γ^1 that is not trivial and satisfies the boundary condition (9) cannot have a trivial x derivative. There must therefore be a point (x, t) at which $u_x \to a \neq 0$ as $\epsilon \to 0$, and if ϵ/κ did not tend to a limit, n_x could not tend to a limit at that point, contrary to (I). But if $\epsilon/\kappa \to 0$, then from (11), u_x would be trivial. Hence, ϵ/κ must tend to a nonzero limit—that is, ord $\kappa = $ ord ϵ; and since (1) to (5) are a homogeneous system, no generality is lost in taking $\kappa(\epsilon) \equiv \epsilon$. The other governing equations can be attacked similarly, and the pattern of argument just sketched thus repeats itself to an extent indicating the emergence of an "asymptotic calculus." A systematic presentation of it has been begun by Olson (1972).

This procedure would determine all of the parameter functions if the number of differential equations were no less than the number of unknown parameter functions. As it is, the analysis divides into two stages, of which the first is characterized by this calculus. For the normal, transverse plasma shock, it determines all the amplitude parameters; in particular, also $\beta(\epsilon) \equiv \epsilon$. Thus the perturbations of average velocity, number density, and magnetic field all have the same amplitude scale.

The second, more arduous stage requires also the solution of limiting differential equations. The four cases to be distinguished cover the possible relations between the amplitude parameter ϵ and the length-scale parameter δ^{-1}. The first is the case that

$$\frac{\epsilon}{\delta^2} \to 0 \qquad \text{as } \epsilon \to 0, \tag{i}$$

which is found (Meyer, 1967c) to imply

$$\tau \equiv \gamma\delta = \delta^3$$

for the time-scale parameter τ^{-1} and

$$b(x, t) \sim b_1 \int_\sigma^\infty Ai(\mu)\, d\mu, \qquad \sigma = \left(\frac{2x^3}{3t}\right)^{1/3}, \tag{12}$$

for the magnetic field perturbation' (Figure IX.5), where Ai denotes the Airy function. (In all cases, the transitions of the number density n and average velocity u have the same form as that of the magnetic field.)

The second case,

$$\text{ord } \delta^2 = \text{ord } \epsilon, \tag{ii}$$

implies (Meyer, 1967c) that ord $\gamma \leq$ ord ϵ and that λ/ϵ tends to a limit, say k. There are thus two subcases. If

$$\text{ord } \gamma = \text{ord } \delta^2 = \text{ord } \epsilon, \tag{iia}$$

the governing equations and scale conditions imply

$$2b_t + (b_{xx} + \tfrac{3}{2}b^2 - kb)_x \qquad \text{is trivial.}$$

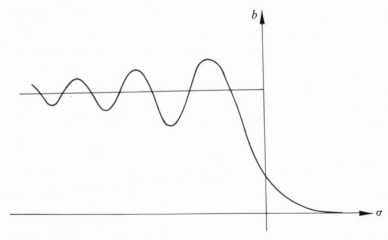

IX.5. Sketch of the variation of b for the case $\epsilon/\delta^2 \nrightarrow 0$ as $\epsilon \to 0$.

This expresses, not so much that the governing equations reduce to

$$2b_t + (b_{xx} + \tfrac{3}{2}b^2 - kb)_x = 0 \tag{13}$$

in the limit (iia), but rather that this is really the differential equation satisfied by the limit of the solution (if there exists any solution satisfying (I) to (III) in the limit (iia)). To obtain limiting differential equations of such a degree of assured reliability is precisely the purpose of the lengthy formulation reported in Sections 5 and 6. Equation (13) will be recognized as the Korteweg-deVries equation (Chapters IV and VIII).

In the other subcase,

$$\text{ord } \gamma < \text{ord } \delta^2 = \text{ord } \epsilon, \tag{iib}$$

it is found that

$$b_{xx} + \tfrac{3}{2}b^2 - kb \text{ is trivial} \tag{14}$$

which implies $k > 0$ and

$$b \sim k \, \text{sech}^2 \, (\tfrac{1}{2}k^{1/2}x), \tag{15}$$

the equation of the solitary wave (Chapters IV and VIII). This corresponds to Morton's (1962) result for the cold limit of steady shock transitions; it satisfies the conditions of Section 6, but of course, represents no shock, since the plasma returns, as $x \to -\infty$, to the original equilibrium (9).

That $\lambda \to 0$, as it does in all these cases, implies by (8) that the Alfvén number U/V_A of such waves is close to unity; in other words, they are magneto-acoustic waves.

Finally, if the Ursell (1953) number

$$\frac{\delta^2}{\epsilon} \to 0 \qquad \text{as } \epsilon \to 0, \tag{iii}$$

it is found that $\lim (\lambda/\epsilon) = k$ exists again, that ord $\gamma \doteq$ ord ϵ and that

$$2b_t + (3b - k)b_x \text{ is trivial.}$$

The limiting equation is therefore of the hyperbolic type discussed in Chapter III. The characteristics, on which $b = const.$, are straight lines of slope $(3b - k)^{-1}$ in the x,t-plane. If $b(x, t)$ decreases with increasing x, at any fixed t, then after a finite time interval a pair of characteristics carrying different values of b must intersect, in contradiction to (I). Hence, case (iii) is possible only if b, u, and n are monotone, non-decreasing functions of x at any fixed t (Figure IX.6) (and this property then persists for all later t).

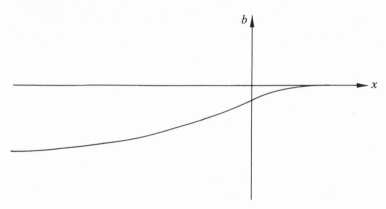

IX.6. Sketch of the variation of b for the case $\delta^2/\epsilon \to 0$.

In this manner, the governing equations and scale conditions were used by Meyer (1967c) to reduce the totality of stretching transformations envisaged in (6) quite rigorously to just three possible ones. Actually, each of the three represents a set of order classes, but a set which is an asymptotic equivalence class in the sense that all its members yield the same first approximation to the solution.

At this point, it should be recalled that the principles of Section 6 do not exhaust the requirements. An eventual approximation to the double limit of small amplitude and near-steadiness in the physically correct order of the limits requires (Section 3) a Kaplun approximation valid for $\tau(\epsilon) = \gamma\delta$ in the left set L_μ of some parameter function $\mu(\epsilon) \in \Omega$. In case (i) above, ord $\tau = $ ord $\delta^3 >$ ord $\epsilon^{3/2}$, which characterizes a right set, not a left set, and hence, (12) cannot be an eventual approximation in the correct sense. Given an arbitrarily small shock strength, (12) can be valid only at degrees of unsteadiness of order larger than $\epsilon^{3/2}$; if steadiness be approached more closely, then the solution cannot continue to be approximated by (12), which therefore can represent only an asymptotic transient.

Actually, (12) is explicit enough to read off a detailed correspondence between the approach to steadiness and the limit $t \to \infty$. As time increases, the wave (12) spreads in proportion to $t^{1/3}$, and (i) must ultimately fail. It may be noted also that case (i) and approximation (12) are the result obtained by the classical approach of letting $\epsilon \to 0$ first and then applying time-asymptotics to the initial-value problem.

In case (iia), ord $\tau =$ ord $(\gamma\delta) =$ ord $\epsilon^{3/2}$, which fails again to represent any left set. Accordingly, the Korteweg-deVries equation (13) can also govern only an asymptotic transient that must give way to a different approximation as steadiness is approached more closely than to the order of $\epsilon^{3/2}$. In case (iib), ord $\tau <$ ord $\epsilon^{3/2}$, so that the resulting approximation could be valid on a left set, but as already noted, it represents no shock. It appears, then, that complete uniqueness has been obtained, after all, by the help of the left-set condition, because only case (iii) is left. Here ord $\tau =$ ord $(\gamma\delta) <$ ord $\epsilon^{3/2}$, so that the approximation can be valid in a left set.

8. Final Waveforms

At the same time, it emerges that near-steadiness is a more ambiguous notion than first supposed (Section 5) and has partially eluded us. A solution corresponding to case (iii) of the preceding section approaches steadiness *locally*, throughout the head of the wave, because it spreads as time increases and thus all derivatives decay, by and by. The approach to local steadiness therefore corresponds again to direct time-asymptotics, and the spreading shows why the condition (iii) is satisfied more strongly with increasing time and hence yields an eventual approximation. But there is no approach to *global* steadiness because the wave spreads at a constant rate (since the characteristics are straight), and accordingly, there is no approach to conservation of mass-flow rate, or momentum-flow rate, or energy-flow rate across the (head of the) wave.

Moreover, Morton's paradox (Section 1) is not really resolved because the monotoneity condition implicit in case (iii) (Section 7) restricts this case to waves lowering the magnetic pressure and number density (Figure IX.6). Apart from mere transients, no possible shock transitions that raise those physical variables have emerged from the analysis.

It therefore becomes necessary to admit that the cases (i) to (iii) of Section 7 are not really exhaustive. The mathematical possibility remains that neither ϵ/δ^2 nor δ^2/ϵ tends to a limit (as noted in Section 2, general parameter functions need not be mutually orderable). Meyer (1967c) and Olson (1972) therefore also checked this fourth case and proved that it is incompatible with the governing equations and scale conditions. This closed the last logical loophole without resolving the paradox.

A reexamination of case (iib)—the only alternative to (iii) that could yield an eventual approximation—was then undertaken. Since $k > 0$, the solitary wave (15) must raise the magnetic pressure, even if only momentarily, and since it is an exceptional solution of (14), a question arises whether even the statement (14) is precise enough. It is derived in the analysis from an equation

$$b_{xx} + \tfrac{3}{2}b^2 - kb = k^2 F, \tag{16}$$

in which the righthand side consists of terms such that F is smooth

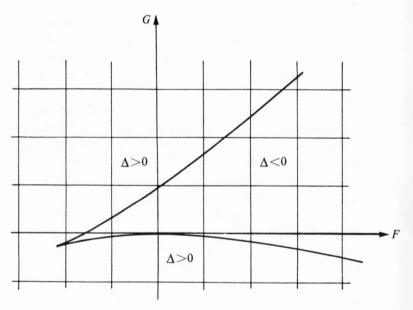

IX.7. Behavior of the discriminant $\Delta(F, G)$.

($\in \Gamma^1$), tends to zero as $x \to +\infty$ for all t, and has trivial derivative F_x. It follows (Section 7) that F must itself be trivial, whence (14).

Before jumping to that conclusion, however, we may integrate (16) to

$$b_x{}^2 = b^2 - b^3 + 2Fb - 2G \equiv C(b; x, t, \epsilon), \tag{17}$$

where $G_x = bF_x$ and $G(\infty, t) = 0$ so that G, like F, is trivial. (The constant k shown positive by (14) may be absorbed into the variables by a linear transformation.) The righthand side of (17) is a near-cubic in b with coefficients F, G, which are *slowly varying* functions of x, and real solutions exist for all x only if the cubic has discriminant $\Delta(F, G) = 0$. Figure IX.7 shows the critical curve $\Delta = 0$ in the F, G-plane (Benjamin and Lighthill, 1954). The initial equilibrium (9) corresponds to the origin, from which the representative point (F, G) can deviate only trivially at any $x < \infty$. But it can be shown (Meyer, 1967c) that the scale conditions imply $\Delta \neq 0$ for $x < \infty$. The precise conclusion in case (iib) is therefore, not (15), but that the scale conditions admit only solutions for which (F, G) moves into the region $\Delta < 0$ (Figure IX.7) by a trivial amount. But any point in $\Delta < 0$, however close to the solitary-wave curve $\Delta = 0$, corresponds to periodic solutions of (17)!

In case (iib), therefore, any solution must eventually approach a *near-periodic wavetrain* (Figure IX.8) raising the magnetic pressure. The first crest differs arbitrarily little from a solitary wave (15) of width of order $\epsilon^{-1/2}$, but further crests of very similar shape follow at a wavelength of order $\epsilon^{-1/2} \log \epsilon$ (Meyer, 1967c). Over a scale of many wavelengths, moreover, the slow change in the parameter F of the train may result in substantial changes of wave shape. But that is beyond the scope of an analysis of the head of the wave.

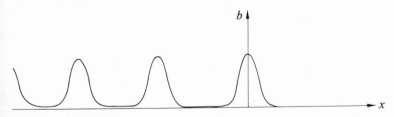

IX.8. Nearly periodic wavetrain that is a possible near-steady transition.

To sum up, there can be at most one weak transition wave with arbitrarily close approach to local steadiness that lowers the magnetic pressure—the monotone wave of case (iii), Section 7—and at most one such wave that raises it, namely a wave starting with near-periodic train just described.

A similar analysis for oblique shocks has been carried out by Olson (1972); the results are markedly different for shocks that differ substantially from normal, transverse ones.

CHAPTER X

Wave Motion in Stratified Fluids

Chia-Shun Yih

If an incompressible fluid at rest has variable density, with the density decreasing with height, or if a compressible fluid has variable entropy, with the entropy increasing with height, a particle of the fluid, when displaced from its original position, will experience a gravitational force that acts to restore it to its original position. This restoring force makes wave motion possible in such a stratified fluid.

Waves can be made not only by an oscillating wave maker or by an obstacle placed in a stream but also as the result of instability. This chapter describes waves in a stratified incompressible fluid at rest, waves created by obstacles in a stratified stream, and waves arising from instability of stratified flows.

1. Waves of Small Amplitude

If x, y, and z are cartesian coordinates, with z measured in the direction of the vertical, the mean density $\bar{\rho}$ (in the absence of waves) of a stratified fluid at rest is a function of z only. The mean pressure $\bar{p}$ is related to $\bar{\rho}$ by the hydrostatic equation

$$\bar{p}_z = -g\bar{\rho}, \tag{1}$$

in which g is the gravitational acceleration. Since three-dimensional infinitesimal waves can be obtained by superposition of two two-

The writing of this chapter, as well as part of the research, has been sponsored by the Office of Naval Research through Grant NRO62-448 to the University of Michigan.

dimensional wavetrains, it is sufficient to consider two-dimensional motion.

Using u and w to denote the velocity components in the x and z directions, respectively, we obtain the linearized equations of motion

$$\bar{\rho}u_t = -p'_x \tag{2}$$

and

$$\bar{\rho}w_t = -p'_z - g\rho', \tag{3}$$

in which t indicates time, and p' and ρ' are the perturbations in pressure and density, respectively. In arriving at (3), the terms involving $\bar{p}$ and $\bar{\rho}$ have been eliminated by the use of (1). The equation of continuity is

$$\rho_t + (\rho u)_x + (\rho w)_z = 0, \tag{4}$$

in which

$$\rho = \bar{\rho} + \rho'$$

is the total density. The equation of incompressibility is

$$\frac{D\rho}{Dt} = 0, \tag{5}$$

in which

$$\frac{D}{Dt} = \frac{\partial}{\partial t} + u\frac{\partial}{\partial x} + w\frac{\partial}{\partial z}.$$

From (4) and (6) we obtain

$$u_x + w_z = 0 \tag{6}$$

as an alternative (and simpler) form of the equation of continuity. Since $\bar{\rho}$ is a function of z only, (5) can be written as

$$\rho'_t + w\bar{\rho}_z = 0, \tag{7}$$

if quadratic terms in the perturbation quantities are neglected. Equations (2), (3), (6), and (7) are the equations on which a linear theory for infinitesimal waves in a stratified incompressible fluid is based.

Equation (6) allows the use of a stream function ψ, in terms of which u and w can be expressed as follows:

$$u = \psi_z, \qquad w = -\psi_x. \tag{8}$$

To study wave motion, we assume ψ to have the factor $\exp(-i\omega t)$, where ω is the "circular frequency" of wave motion, or 2π times the frequency. Elimination of p' and ρ' among (2), (3), and (7) and use of (8) produce

$$(\omega^2 \bar{\rho} + g\bar{\rho}') \frac{\partial^2 \psi}{\partial x^2} + \omega^2 \frac{\partial}{\partial z} \left(\bar{\rho} \frac{\partial \psi}{\partial z} \right) = 0, \tag{9}$$

in which

$$\bar{\rho}' = \frac{d\bar{\rho}}{dz}.$$

It is evident that at any point (9) is of the elliptic or hyperbolic type according as

$$\omega^2 > -\frac{g\bar{\rho}'}{\bar{\rho}} \quad \text{or} \quad \omega^2 < \frac{-g\bar{\rho}'}{\bar{\rho}}. \tag{10}$$

If

$$\omega^2 > g \left(-\frac{\bar{\rho}'}{\bar{\rho}} \right)_{\max}, \tag{11}$$

then (9) is elliptic everywhere and wave motion is impossible. The square root of the righthand side of (11) is called the Brunt-Väisälä frequency. If a body oscillates in a stratified fluid with a frequency higher than the Brunt-Väisälä frequency, it creates no waves in the fluid, which oscillates with the body in much the same way as a homogeneous fluid without a free surface. If the second of the two inequalities in (10) holds, (9) possesses two characteristic lines passing through the point in question, and wave motion is possible there. Of course the first inequality in (10) may hold in certain regions of the fluid, and the second inequality holds elsewhere. In that case (9) is of the mixed type, and its solution is in general difficult. (See Görtler, 1943, and the more recent work of Mowbray and Rarity, 1967, for further information in this connection.)

1a. The Differential Equation for Normal Modes

For the two-dimensional waves propagating in the x direction, we may assume

$$\psi = f(z) \exp[ik(x - ct)], \tag{12}$$

in which k is the wavenumber defined by

$$k = \frac{2\pi}{\lambda},$$

λ being the wavelength, and c is the phase velocity.

Equations (8) and (12) give

$$u = f'(z) \exp [ik(x - ct)], \tag{13}$$

$$w = -ikf(z) \exp [ik(x - ct)], \tag{14}$$

in which $f'(z) = df(z)/dz$. It is understood that the real parts of the right-hand sides of (13) and (14) are to be used. Examination of (7) shows that ρ' has the same exponential factor as ψ; hence,

$$ikc\rho' = w \frac{d\bar{\rho}}{dz}. \tag{15}$$

Equations (13) to (15) show how to obtain u, w, and ρ' from $f(z)$. To obtain the equation governing f, we can substitute (13) and (15) into (2) and (3) and eliminate p', or, what amounts to the same thing, we can substitute (12) into (9). The result is

$$(\bar{\rho}f')' - \left(k^2 \bar{\rho} + \frac{g\bar{\rho}'}{c^2} \right) f = 0 \tag{16}$$

or

$$(\bar{\rho}f')' - k^2 \left(\bar{\rho} + \frac{g\bar{\rho}'}{\omega^2} \right) f = 0, \tag{17}$$

where the primes indicate differentiation with respect to z.

1b. Boundary Conditions

Let the depth of the fluid be d. If the lower and upper boundaries are rigid planes, w must vanish at these boundaries, and the boundary conditions are

$$f(0) = 0 \quad \text{and} \quad f(d) = 0. \tag{18}$$

If there is a discontinuity in density $\bar{\rho}$ at some value of z, at which the density $\bar{\rho}$ jumps from $\bar{\rho}_l$ to $\bar{\rho}_u$ ($< \bar{\rho}_l$), the interfacial condition at this discontinuity can be obtained by integrating (16) in the Stieltjes sense (allowing infinite values of the integrand) across the discontinuity. Inspection of (16) shows that there is a jump in $\rho f'$ because there is one in $\bar{\rho}'$, since f is continuous. Thus integration across the discontinuity (from $z_0 - \epsilon$ to $z_0 + \epsilon$, z_0 being the point of density discontinuity) gives

$$(\bar{\rho}f')_u - (\bar{\rho}f')_l + \frac{g}{c^2}(\bar{\rho}_l - \bar{\rho}_u)f = 0, \tag{19}$$

which is the interfacial condition. Note that while f is continuous at $z = z_0$, f' may not be and in fact is not. At a free surface (if there is one), $\bar{\rho}_u = 0$. Therefore, if the lower boundary is a rigid plane and the upper surface of the fluid is free, the boundary conditions are

$$f(0) = 0 \quad \text{and} \quad f'(d) = \frac{g}{c^2}f(d), \tag{20}$$

the second of which follows from (19).

From (17) we can also see the implication of the inequality (11). For (11) to hold, $\bar{\rho}$ must not have any discontinuity, and, in particular, the fluid must not have a free surface. The boundary conditions are then given by (18). Multiplying (17) by f and integrating, using (18) whenever necessary, we obtain

$$-\int_0^d \bar{\rho}f'^2 \, dz - k^2 \int_0^d \left(\bar{\rho} + \frac{g\bar{\rho}'}{\omega^2}\right)f^2 \, dz = 0,$$

which is quite impossible if $k^2 \geq 0$ and (11) holds. Hence wave motion is impossible if the frequency is greater than the Brunt-Väisälä frequency, as we have seen already from the viewpoint of the type of (9).

1c. Proof that ω^2 is Real

Whether $\bar{\rho}'$ is positive, negative, or partly positive and partly negative, whether there are discontinuities in $\bar{\rho}$ or not, ω^2 is real. To show this we note that, with (12) substituted into (9), we obtain

$$\omega^2 [(\bar{\rho}f')' - k^2\bar{\rho}f] - k^2 g\bar{\rho}'f = 0. \tag{21}$$

We note first of all that ω cannot be zero, for otherwise (21) would show that $f = 0$, since $\bar{\rho}'$ is not identically equal to zero. It was only after division of this result by ω^2 that we obtained (17). We prefer to use (21) for the proof of the reality of ω^2.

To show that ω^2 is real for the case of no density discontinuities, multiply (21) by f^*, the complex conjugate of f, and integrate, using (18) whenever necessary, to obtain

$$\omega^2 \int_0^d [\bar\rho |f'|^2 + k^2 \bar\rho |f|^2]\, dz + k^2 g \int_0^d \bar\rho' |f|^2\, dz = 0. \tag{22}$$

Equation (22) shows that ω^2 is real even if $\bar\rho'$ changes sign in $0 \le z \le d$. If the upper surface is free, the term

$$\bar\rho(d)\, \omega^2\, \frac{g}{c^2} |f|^2 = \bar\rho(d)\, k^2 g\, |f|^2$$

is subtracted from the lefthand side of (22). It is clear that if there are density discontinuities (including the free surface, if there is one), upon the use of (19) (and (20) if there is a free surface) one need only consider the last integral in (22) to be in the Stieltjes sense. That is to say, if there are N continuously stratified layers, one need only replace the last integral in (22) by

$$\sum_{n=1}^{N} \int_{\cdot}^{\cdot} \bar\rho' |f|^2\, dy - \sum_{n=1}^{M} (\Delta\bar\rho |f|^2)_n,$$

in which the integrals are performed over the N layers, exclusive of the discontinuities, $\Delta\bar\rho$ is the jump in the mean density $\bar\rho$, and the subscript n means the value is taken at the nth discontinuity. The value of M is N if there is a free surface and $N-1$ if there is not. By exactly the same argument as that for the case of no density discontinuities, we conclude that ω^2 is real.

Whether there are density discontinuities or not, if the density always decreases as z increases (and $\bar\rho_l > \bar\rho_u$ at every discontinuity if there are density discontinuities), ω^2 must be positive as can be seen from (22), in which the last integral is in the Stieltjes sense if there are density discontinuities. If the density always increases with height, with $\bar\rho_u > \bar\rho_l$ at all density discontinuities if any, then (22) shows that ω^2 is negative. A negative ω^2 means two imaginary values of ω, one of which makes $\exp(-i\omega t)$ an exponentially increasing function of time, so that the fluid is unstable. If $\bar\rho$ increases with height in part of the fluid and decreases with height elsewhere, ω^2 can be positive or negative.

1d. Existence of Eigenvalues

If there are no internal density discontinuities, the Sturm-Liouville theory as extended by Bôcher (1917) guarantees the existence of eigenvalues of ω^2 whether or not the upper surface is free. Indeed, the following oscillation theorem of Sturm not only states this existence but also says something important about the eigenfunctions (Bôcher, 1917, pages 66-67):

THE STURM-BÔCHER THEOREM. *Consider the system*

$$\left. \begin{array}{l} \dfrac{d}{dz}\,(Kf') - Gf = 0, \\[2mm] \alpha'f(a) - \alpha f'(a) = 0, \\[1mm] \beta'f(b) + \beta f'(b) = 0, \end{array} \right\}$$

in which K is always positive, and K and G are functions of z and the parameter λ, *which do not increase as* λ *increases from* Λ_1 *to* Λ_2 ($> \Lambda_1$), *and* α, α', β, *and* β' *are functions of* λ. *If* $\beta = 0$, *or* $\beta \neq 0$ *and* $K(b)\beta'/\beta$ *is a decreasing function of* λ, *and if, in addition,*

$$\lim_{\lambda \to \Lambda_2} \left(\frac{-\max G}{\max K} \right) = +\infty$$

and

$$\lim_{\lambda \to \Lambda_1} \left(\frac{-\min G}{\min K} \right) = -\infty,$$

then the system has an infinite number of eigenvalues λ_0, λ_1, $\ldots$, *between* Λ_1 *and* Λ_2, *in ascending order of magnitude. Each of the eigenfunctions* f_0, f_1, $\ldots$, *which are solutions of the system for* $\lambda = \lambda_0$, λ_1, $\ldots$, *has a number of zeros in* $a < z < b$ *exactly equal to its respective index.*

We note that the system consisting of (16) and (18) or (20), or of (17) and (18) or (20), satisfies every requirement of the Sturm-Liouville system stated in the theorem above if we identify λ with c^{-2} or ω^{-2}. Note also that if the boundary conditions are given by (18), then there are $n + 2$ zeros in $0 \leq z \leq d$ for $\lambda = \lambda_n$. (Here $a = 0$ and $b = d$.) If the boundary conditions are given by (20), then there are $n + 1$ zeros in $0 \leq z \leq d$

for $\lambda = \lambda_n$. If $\bar\rho'$ is negative throughout, it can be readily verified that, with $\lambda = c^{-2}$, for (16) $\Lambda_2 = +\infty$ $(c^2 = +0)$ and $\Lambda_1 = -\infty$ $(c^2 = -0)$. Similarly, with $\lambda = \omega^{-2}$ in (17), $\Lambda_2 = +\infty$ $(\omega^2 = +0)$ and $\Lambda_1 = -\infty$ $(\omega^2 = -0)$. Since we know that if $\bar\rho'$ is negative throughout, c^2 and ω^2 are both real and positive, and for any k^2 the value of ω^2 (and also of c^2) is bounded, all the eigenvalues of λ are between some positive value and $+\infty$ —that is, all the eigenvalues of ω^2 or c^2 are between zero and some finite, positive, upper bound.

If $\bar\rho'$ is uniformly negative where $\bar\rho$ is continuous, and at all density discontinuities $\bar\rho_u < \bar\rho_l$, the eigenvalues for c^2 (or ω^2) have been shown to exist (see Yih, 1965, pages 48-52). The part of the Sturm-Bôcher theorem that relates the index of the eigenvalues to the zeros of the eigenfunctions is still true even if $\bar\rho$ has discontinuities. This can be seen in the following way. First let the upper surface be rigid. For $\lambda = 0$ (or $c^2 = \infty$), obviously $f(z)$ increases with z as can be seen from (16) or (17) if $f'(0)$ is positive. Hence $f(d)$ is not zero. (If $f'(0)$ is negative, then $f(z)$ decreases as z increases, and we have the same result.) Since $c^2 = \infty$, this is true whether or not there are discontinuities in $\bar\rho$, as a glance at the interfacial conditions (19) will make clear. As we increase λ (or decrease c^2), $f(d)$ will vanish when the first eigenvalue λ_0 for fixed upper surface is reached. The existence of this λ_0 is known (Yih, 1965, pages 48-52). As subsequent eigenvalues λ_1, λ_2, $\ldots$ are passed, the number of internal zeros is increased by 1 at each such passing. For otherwise there are two possibilities:

   (i) Zeros can be created or destroyed internally.
   (ii) The passing of λ_m (for any integral m) does not increase the number of internal zeros by 1.

If (i) were possible, there would be a stage at which two (or more) zeros coalesce at $z = z_a$, for instance, since the positions of the zeros are continuous functions of λ for given $\bar\rho$ and k^2. But a double zero for $f(z)$ at $z = z_a$ would mean the vanishing of $f(z)$ identically everywhere, and therefore (i) cannot happen. If (ii) were possible, $f(d)$ would have a double zero at $\lambda = \lambda_m$ (for otherwise $f(d)$ would change sign as λ_m is crossed), and it can be shown that double eigenvalues are not possible. If the upper surface is free, it can be shown that $f'(d)/f(d)$ decreases from a positive value to negative infinity as λ increases from zero toward λ_0 (the first eigenvalue for λ for the case of rigid upper surface) so that at some $\lambda < \lambda_0$ the free-surface condition (20) is satisfied. Similarly, if λ_{n-1} denotes the nth eigenvalue of λ for the rigid-upper-surface case, as λ varies from λ_{n-1} to λ_n, $f'(d)/f(d)$ decreases from positive infinity to negative infinity, so that (20) is satisfied at some λ between λ_{n-1} and λ_n. Furthermore, since zeros of the eigenfunction can be neither created nor destroyed internally, and as each λ_m is passed the internal zeros are increased by 1, that part of the Sturm-Bôcher theorem concerning the index of eigenvalues and the internal zeros of eigenfunctions remains true even if there are internal discontinuities in $\bar{\rho}$ and a free surface.

It is evident that if $\bar{\rho}'$ is everywhere positive and $\bar{\rho}_l < \bar{\rho}_u$ at every density discontinuity (if any), the existence of the eigenvalues of c^2, necessarily negative, and the relationship between the index of the eigenvalue and the internal zeros of the eigenfunction as stated by the Sturm-Bôcher theorem follow immediately from that theorem and the discussion in the preceding paragraph if we simply take $\lambda = -c^{-2}$. To each eigenvalue of c^2 at constant k^2 corresponds an eigenvalue of ω^2, of course; hence we need not discuss the existence of the eigenvalues of ω^2 independently.

If $\bar{\rho}'$ is positive in part of the fluid and negative elsewhere, by the use of the comparison theorems of Sturm to be presented in the next section we can show that there are two infinite series of eigenvalues of c^2, one positive and one negative, both tending to zero as a limit point, and for each series f_n ($n = 0, 1, \ldots$) corresponding to λ_n has exactly n internal zeros. This situation is unchanged when there are density discontinuities, for some of which $\bar{\rho}_l < \bar{\rho}_u$, and for others of which $\bar{\rho}_u < \bar{\rho}_l$.

1e. The Variation of ω^2 and c^2 with k^2

Like gravity waves in a homogeneous fluid, gravity waves in a stratified fluid are also dispersive. That is to say, their velocity is a function of k. That this must be so can be seen from (16) or (17) and their boundary conditions. Without detailed calculations, which are in any case impossible unless $\bar{\rho}$ is specified, we can show that c^2 always decreases and ω^2 always increases as k^2 increases. For this purpose we need the comparison theorems of Sturm.

Consider the two systems (Bôcher, 1917, pp. 58-63)

$$(K_1f'_1)' - G_1f_1 = 0, \qquad f(a) = \alpha_1, \quad f'(a) = \alpha'_1,$$

and

$$(K_2f'_2)' - G_2f_2 = 0, \qquad f(a) = \alpha_2, \quad f'(a) = \alpha'_2,$$

where K_1, K_2, G_1, and G_2 are functions of z (and possibly of a parameter also), and

$$K_1 \geq K_2 > 0, \qquad G_1 \geq G_2.$$

It is assumed that

$$|\alpha_1| + |\alpha'_1| \neq 0, \qquad |\alpha_2| + |\alpha'_2| \neq 0,$$

that the equalities $K_1 = K_2$ and $G_1 = G_2$ do not hold in any part of the interval $[a, b]$, and that $G_1 = 0 = G_2$ does not hold in any part of the interval. Furthermore, if $\alpha_1 \neq 0$, it is assumed that $\alpha_2 \neq 0$ and

$$\frac{K_1(a)\alpha'_1}{\alpha_1} \geq \frac{K_2(a)\alpha'_2}{\alpha_2}.$$

If $\alpha_1 = 0$, no supplementary assumption need be made. Under the preceding assumptions, we have the following comparison theorems of Sturm:

STURM'S FIRST COMPARISON THEOREM. *If f_1 has a certain number of zeros in the interval $a < z \leq b$, f_2 must have at least as many zeros in this interval, and if $z_1, z_2, z_3, \ldots$ are the zeros of f_1 in ascending order, and $z'_1, z'_2, z'_3, \ldots$ are those of f_2, then*

$$z'_i < z_i$$

for all values of i corresponding to a zero of f_1 and f_2.

STURM'S SECOND COMPARISON THEOREM. *If, in addition to the assumptions stated in this paragraph, it is assumed that $f_1(b) \neq 0$ and $f_2(b) \neq 0$, then*

$$\frac{K_1(b)f'_1(b)}{f_1(b)} > \frac{K_2(b)f'_2(b)}{f_2(b)},$$

provided that f_1 and f_2 have the same number of zeros in $a < z \leq b$.

We are now in a position to see how c^2 or ω^2 varies with k^2. Conside first a stably stratified fluid with no internal density discontinuities. If the upper boundary is fixed, we can conclude from Sturm's first comparison theorem that for the same mode—that is, for a fixed number of zeros of the eigenfunction in $0 < z \leq d$—c^2 decreases as k^2 increases. For if $k_1{}^2 > k_2{}^2$ and $c_1{}^2$ and $c_1{}^2 \geq c_2{}^2$,

$$G_1 \equiv k_1{}^2 \bar{\rho} + \frac{g\bar{\rho}'}{c_1{}^2} > k_2{}^2 \bar{\rho} + \frac{g\bar{\rho}'}{c_2{}^2} \equiv G_2 \,,$$

and if $z = d$ is the nth zero of f_1, it cannot be the nth zero of f_2, which must occur at $z < d$. Hence, either f_2 does not vanish at $z = d$, violating the boundary condition there, or f_2 has at least one more zero than f_1 in $0 < z \leq d$. Hence, if $k_1{}^2 > k_2{}^2$, we must have $c_1{}^2 < c_2{}^2$. This result was first obtained by Groen (1948).

If the upper surface is free, use of Sturm's second comparison theorem leads to the same result. (See Yih, 1965, pages 32-33.)

If there are density discontinuities but the density never increases continuously or discontinuously as z increases, the same result can be used upon repeated applications of the comparison theorems of Sturm. But this application is tedious and inconvenient for exposition. Hence we shall present a new and more elegant proof of the theorem that c^2 decreases as k^2 increases, particularly since it can be used to show that ω^2 *increases* with k^2. Consider (16) with (18) and (19) or with (19) and (20). Let k^2 increase by dk^2 and c^2 by dc^2, and let the corresponding variation in f be ϵ. Then ϵ satisfies

$$(\bar{\rho}\epsilon')' - \left(k^2\bar{\rho} + \frac{g\bar{\rho}'}{c^2}\right)\epsilon = (dk^2)\bar{\rho}f - \frac{g\bar{\rho}'}{c^4}(dc^2)f. \tag{23}$$

The boundary condition for ϵ is $\epsilon = 0$ at a rigid surface, and

$$\epsilon' = \frac{g}{c^2}\epsilon - \frac{g(dc^2)}{c^4}f$$

at a free surface. The conditions on ϵ at an internal surface of density discontinuity can be similarly obtained. We can now multiply (16) by ϵ and (23) by f, integrate (layer by layer if there are density discontinuities) between zero and d, and apply the boundary and interfacial conditions on f and ϵ. The difference of the two equations so obtained is, with the last integral understood in the Stieltjes sense if there are any density discontinuities,

$$0 = (dk^2) \int_0^d \bar{\rho} f^2 \, dz - \frac{g(dc^2)}{c^4} \int_0^{d+} \bar{\rho}' f^2 \, dz. \tag{24}$$

Since the first integral in (24) is positive and the second negative, (24) means that

$$\frac{dk^2}{dc^2} < 0. \tag{25}$$

Note that since we have obtained (25) by perturbation, which does not change the mode, (25) is automatically for the same mode only. Here lies the simplicity of this approach.

We can also show that for the same mode

$$\frac{d\omega^2}{dk^2} > 0. \tag{26}$$

All that is necessary is to replace (16) by (17), (23) by

$$(\bar{\rho}\epsilon')' - k^2\left(\bar{\rho} + \frac{g\bar{\rho}'}{\omega^2}\right)\epsilon = (dk^2)\left(\bar{\rho} + \frac{g\bar{\rho}'}{\omega^2}\right)f - \frac{k^2 g\bar{\rho}'}{\omega^4}(d\omega^2)f,$$

and to note that multiplication of (17) by f and integration shows that

$$\int_0^d \left(\bar{\rho} + \frac{g\bar{\rho}'}{\omega^2}\right)f^2 \, dz < 0.$$

The rest follows as before, and (26) results.

We can choose k and c to be positive. Then ω is positive, and (25) and (26) can be replaced by

$$\frac{dc}{dk} < 0, \tag{27}$$

$$\frac{d\omega}{dk} > 0. \tag{28}$$

Since $d\omega/dk$ is the group velocity C, (28) shows that C is in the same direction as c, and (27) shows that

$$C = \frac{d\omega}{dk} = \frac{d(kc)}{dk} = c + k\frac{dk}{dc} < c.$$

For gravity waves of a homogeneous liquid, $C < c$ has the significance that there are no upstream waves ahead of an obstacle placed in a stream of uniform velocity or moving in a stratified fluid at rest.

We have just discussed the propagation of normal modes in the x direction. The direction of group velocity in a two-dimensional wave motion propagating in the x-z plane (not merely in the x direction) can have some peculiar and strange behavior. (See Mowbray and Rarity, 1967, and the brief discussion in Yih, 1969b.)

2. Waves of Finite Amplitude

We shall first consider two-dimensional gravity waves created by an obstacle placed in a stratified fluid in steady flow. The equations of motion for steady two-dimensional flows are

$$\rho(uu_x + wu_z) = -p_x, \tag{29}$$

and

$$\rho(uw_x + ww_z) = -p_z - g\rho, \tag{30}$$

in which ρ is the density and p the pressure, the other symbols retaining their meanings given in Section 1. The equation of incompressibility is

$$u\rho_x + w\rho_z = 0 \tag{31}$$

by virtue of which the equation of continuity has the form (6). If we define (Yih, 1958)

$$(u', w') = \sqrt{\frac{\rho}{\rho_0}} (u, w),$$

in which ρ_0 is a constant reference density, because of (31), the equations (29), (30), and (6) can be written as

$$\rho_0(u'u'_x + w'u'_z) = -p_x, \tag{32}$$

$$\rho_0(u'w'_x + w'w'_z) = -p_z - g\rho, \tag{33}$$

and

$$u'_x + w'_z = 0. \tag{34}$$

Equations (32) and (33) are simpler than (29) and (30) because ρ_0 is constant, whereas the ρ on the lefthand sides of (29) and (30) is variable. We shall therefore work with (32), (33), and (34).

Equation (34) allows the use of the stream function ψ', in terms of which

$$u' = \psi'_z, \qquad w' = -\psi'_x,$$

and the y component of the vorticity of the associated flow (with velocity components u' and w') is

$$\eta' = u'_z - w'_x = \nabla^2 \psi',$$

where ∇^2 is the two-dimensional Laplacian. We can write (32) and (33) as

$$\rho_0 \eta' \, \psi'_x = \left[p + \frac{\rho_0(u'^2 + w'^2)}{2} \right]_x, \tag{35}$$

$$\rho_0 \eta' \, \psi'_z = \left[p + \frac{\rho_0(u'^2 + w'^2)}{2} \right]_z + g\rho. \tag{36}$$

Multiplying (35) by dx and (36) by dz and adding the results, we obtain

$$\rho_0 \eta' \, d\psi' = d\left[p + \frac{\rho(u^2 + w^2)}{2} \right] + g\rho \, dz = dH - gz \, d\rho, \tag{37}$$

in which

$$H = p + \frac{\rho(u^2 + w^2)}{2} + g\rho z$$

is the Bernoulli function, which does not vary along a streamline and therefore, like ρ, is a function of ψ' alone. Equation (37) can thus be written as

$$\nabla^2 \psi' + \frac{gz}{\rho_0} \frac{d\rho}{d\psi'} = \frac{1}{\rho_0} \frac{dH}{d\psi'} = h(\psi'), \tag{38}$$

(Yih, 1958) which is a simpler form of the corresponding equation in terms of Ψ [stream function for the actual flow related to u and w by (8)] obtained first by Madame Dubreil-Jacotin (1935) and then independently by Long (1953):

$$\nabla^2 \psi + \frac{1}{\rho} \frac{d\rho}{d\psi} \left(\frac{\psi_x^2 + \psi_z^2}{2} \right) + gz = f(\psi). \tag{39}$$

Because of its simplicity, we shall use (38) instead of (39).

It is appropriate to note here that the analogous equations governing steady three-dimensional finite-amplitude motion of a stratified fluid are also known (Yih, 1967) but are difficult to solve.

2a. Lee Waves in a Stratified Stream

Given a density and a velocity distribution upstream, ρ and H in (38) can be evaluated as functions of ψ'. Thus (38) is in general nonlinear. Linear forms of (38) can, however, be obtained simply by demanding $d\rho/d\psi'$ and $h(\psi')$ to be linear in ψ'. One can then determine the corresponding upstream conditions. Fortunately, the classes of upstream conditions so obtained are sufficiently realistic, numerous, and adjustable to approximate actual conditions far upstream. We shall limit our discussion to lee waves created by an obstacle placed in a stratified stream for which (38) has a particularly simple linear form. Note that a linear form for (38) does not result from a linearization and does not require that the disturbances be small.

Consider a flow in the region $-\infty \le x \le 0$, $0 \le z \le d$, with

$$u' = U' = const. \quad \text{and} \quad \rho = \rho_0 - \frac{\rho_0 - \rho_1}{d} z \text{ at } x = -\infty. \tag{40}$$

If we use the dimensionless variables

$$\xi = \frac{x}{d}, \qquad \eta = \frac{z}{d}, \qquad \Psi = \frac{\psi'}{U'd},$$

then at $x = -\infty$ we have

$$\Psi = \eta, \qquad \nabla^2\Psi = 0, \quad \text{and} \quad \frac{d\rho}{d\psi'} = \frac{\rho_1 - \rho_0}{U'd},$$

and we find $h(\psi')$ to be proportional to ψ' or Ψ, and (38) to have the form

$$\Psi_{\xi\xi} + \Psi_{\eta\eta} - F^{-2}\eta = -F^{-2}\Psi, \tag{41}$$

where

$$F^2 = \frac{U'^2}{g'd}, \qquad g' = g\frac{\rho_0 - \rho_1}{\rho_0}.$$

If there is a barrier in the stream, it will disturb that stream, and if the Froude number F is sufficiently low, there will be waves in the lee of the barrier. To represent a barrier, we shall use the method of singularities described below.

Let a line of singularities be situated at $x = 0$. We shall find the solution of (41) for $x < 0$, which we call Ψ_-, and the solution of (41) for $x > 0$, which we call Ψ_+. Then we shall match Ψ_- to Ψ_+ at $x = 0$, taking into account the presence of the singularities. The resulting solution then describes the flow past the barrier created by the line of singularities. As we shall see, we can use more than one line of singularities.

Taking into account the experimentally observed absence of waves upstream from the barrier (which can be explained in the case of infinitesimal waves but will be assumed here), we have (Yih, 1960)

$$\Psi_- = \eta + \sum_{N+1}^{\infty} A_n e^{a_n \xi} \sin n\pi\eta \qquad \text{for } \xi < 0,$$

$$\Psi_+ = \eta + \sum_{1}^{N} (B_n \cos a_n \xi + C_n \sin a_n \xi) \sin n\pi\eta \qquad (42)$$

$$+ \sum_{N+1}^{\infty} D_n e^{-a_n \xi} \sin n\pi\eta \qquad \text{for } \xi > 0,$$

in which

$$a_n = |n^2\pi^2 - F^{-2}|^{1/2} \qquad (43)$$

and N is defined by

$$N^2\pi^2 < F^{-2} < (N+1)^2\pi^2. \qquad (44)$$

Note that each term in (42) is a solution of (41), that the term η corresponds to the undisturbed flow, and that the solution for Ψ_- contains no wave terms. The number of lee-wave components is determined by (44), and the wavelengths are determined by (43).

We can demand that

$$\Psi_- = \Psi_+ \qquad \text{at } \xi = 0, \qquad (45)$$

$$\frac{\partial \Psi_-}{\partial \xi} - \frac{\partial \Psi_+}{\partial \xi} = f(\eta) \qquad \text{at } \xi = 0, \qquad (46)$$

where

$$f(\eta) = 0 \qquad \text{for } d > \eta > a$$

and $f(\eta)$ is specified for $\eta \le a$. Equation (45) means that there are no sources and sinks along the line $\xi = 0$, and (46) means that there is a vortex sheet at $\xi = 0$ for $\eta \le a$. Condition (45) demands that $A_n = D_n$ for $n > N$, and $B_n = 0$ for $n \le N$, and (39) demands that

$$a_n(A_n + D_n) = 2 \int_0^1 f(\eta) \sin n\pi\eta \, d\eta \qquad \text{for } n > N$$

and

$$a_n C_n = -2 \int_0^1 f(\eta) \sin n\pi\eta \, d\eta \quad \text{for } n \le N. \qquad (47)$$

Thus, once $f(\eta)$ is specified, all the coefficients in (42) are known. Equation (47) is especially significant, for it states that the C_n's, which are the amplitudes of the lee-wave components, depend on only the first N Fourier coefficients of $f(\eta)$ and not on the rest of them. That is to say, they do not depend on the complete details of $f(\eta)$, hence do not depend on the complete details of the barrier. Also, the form of Ψ'_+ in (42) shows that there are N lee-wave components, since $B_n = 0$.

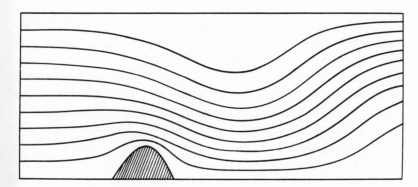

X.1. Pattern of a stratified flow with waves in the lee of a barrier. One lee-wave component.

Figure X.1 shows the flow pattern at $F = 3/4\pi$, with a barrier created by the singularity distribution

$$f(\eta) = -10 \sin 5\pi\eta \qquad \text{for } 0 \le \eta \le 0.2,$$
$$f(\eta) = 0 \qquad\qquad \text{for } 0.2 \le \eta \le 1.$$

There is only one lee-wave component.

Instead of demanding (45) and (46), we can demand

$$\Psi_- - \Psi_+ = f(\eta) \qquad \text{at } \xi = 0$$

and

$$\frac{\partial \Psi_-}{\partial \xi} = \frac{\partial \Psi_+}{\partial \xi} \qquad \text{at } \xi = 0.$$

The line singularity is then a distribution of sources and sinks. For a closed barrier, the integral of $f(\eta)$ from zero to a must be zero. The coefficients in (42) are then given by

$$
\begin{aligned}
A_n &= -D_n & \text{for } n > N, \\
C_n &= 0 & \text{for } n \le N, \\
A_n - D_n &= 2 \int_0^1 f(\eta) \sin n\pi\eta \, d\eta & \text{for } n > N,
\end{aligned}
$$

and

$$B_n = -2 \int_0^1 f(\eta) \sin n\pi\eta \, d\eta \qquad \text{for } n \le N.$$

We can impose singularities at other values of ξ than $\xi = 0$, whether we use vorticity distributions or source-sink distributions, or both. For a closed body, the sum of the line integrals of the source-sink distributions must be zero. Solutions of (41) are also given by Long (1953), Drazin and Moore (1967), and Miles (1968a, b). For discussion of these works in connection with upstream influence and the phenomenon of blocking see the survey article of Yih (1969b). Here we remark only that although upstream influence always exists, a solution of (41) for specified upstream conditions is not necessarily without significance, for these conditions can be considered as the result of placing a barrier in a flow originally different from the upstream flow that results. Solutions for high barriers at low Froude numbers and with very complicated flow patterns and regions of closed streamlines may not be realistic, however, since blocking can occur in those cases to change the upstream conditions from those assumed in the solutions.

We conclude this section by noting that other classes of finite-amplitude wave motions are known for stratified fluids.

Solitary waves in an incompressible fluid with an exponential distribution in density were studied by Long (1965), and solitary and cnoida waves in a stratified incompressible fluid by Benjamin (1966). In these studies, the amplitude of motion in the vertical direction is taken to be that for periodic waves of long wavelength, and a stretching (or rather contraction) of the horizontal coordinate in the direction of wave propagation gives the equation that determines the waveform after the dynamical equations have been satisfied at the required order of approximation.

A class of solutions according to the shallow-water theory for gravity waves in a stratified fluid is known (Yih, 1969a). In addition, Gerstner's solution for finite-amplitude waves, found to remain valid for a stratified fluid by Madame Dubreil-Jacotin (1932), has been discovered to represent edge waves in a stratified fluid also (Yih, 1966).

3. Stability

Since waves may be generated by instability of stratified flows, it is appropriate to discuss the question of stability. But we shall limit our discussion to general results.

Since the stability or instability of three-dimensional disturbances in a stratified flow can be deduced from the stability or instability of two-dimensional disturbances in a flow with reduced mean velocity (Yih, 1955), we shall study two-dimensional disturbances only.

For convenience we shall henceforth use y for the vertical coordinate instead of z, and v for the vertical component of the velocity instead of w. If the mean velocity in the x direction denoted by U is a continuous function of y, u and v denote the components of the perturbation velocity in the directions of increasing x and y, respectively, and $\bar{\rho}$ still denotes the mean density, the linearized equations of motion are

$$\bar{\rho}(u_t + Uu_x + U'v) = -p_x \tag{48}$$

and

$$\bar{\rho}(v_t + Uv_x) = -p_y - g\rho, \tag{49}$$

in which p is now the pressure perturbation and ρ the density perturbation and the prime denotes differentiation with respect to y. The equation of continuity

$$u_x + v_y = 0$$

again permits the use of the stream function ψ in terms of which

$$u = \psi_y, \qquad v = -\psi_x. \tag{50}$$

The linearized equation of incompressibility is

$$\rho_t + U\rho_x + v\bar{\rho}' = 0. \tag{51}$$

If η is the vertical displacement of a line of constant density from its mean position, the kinematic equation for that line is

$$\eta_t + U\eta_x = v = -\psi_x. \tag{52}$$

All perturbation quantities are assumed to have the factor exp $ik(x - ct)$, which completely represents their dependence on x and t. Hence (52) and (50) give

$$\psi = -(U - c)\eta, \qquad u = -[(U - c)\eta]' v = ik(U - c)\eta, \tag{53}$$

and (48) and (51) give

$$p = \bar{\rho}(U - c)^2\eta' \quad \text{and} \quad \rho = -\bar{\rho}'\eta. \tag{54}$$

If $F(y)$ is the amplitude of η—that is, if

$$\eta(x, y, t) = F(y) \exp ik(x - ct)$$

—substitution of (53) and (54) into (49) produces

$$[\bar{\rho}(U - c)^2 F']' + \bar{\rho}[\beta g - k^2(U - c)^2]F = 0, \tag{55}$$

where

$$\beta = -\frac{\bar{\rho}'}{\bar{\rho}} > 0.$$

Equation (55) and the boundary conditions constitute the system governing stability. The eigenvalue c may be complex. Let $c = c_r + ic_i$. If c_i is positive, the flow is unstable. It can be easily verified that the boundary conditions (whether the upper surface is free or not, and whether internal density discontinuities exist), like (55), do not involve the imaginary number i explicitly. Hence, if c is an eigenvalue, so is its complex conjugate c^*. Hence, if c is complex, there is an eigenvalue of c with positive c_i, and the flow is unstable.

Instead of using (55), which is an equation for η, we can obtain an equation for ψ by writing

$$\psi = f(y) \exp ik(x - ct) \tag{56}$$

and working with (50), (48), and (49). Elimination of p between (48) and (49) gives

$$(\bar{\rho}f')' + \left[\frac{(\bar{\rho}U')'}{c - U} - k^2\bar{\rho} - \frac{g\bar{\rho}'}{(c - U)^2}\right]f = 0. \tag{57}$$

Of course, we can also use (56) and the first equation of (53) to obtain

$$f(y) = -(U - c)F(y)$$

and substitute this in (55) to obtain (57).

If the upper surface is fixed, the boundary conditions for $F(y)$ and $f(y)$ are, respectively,

$$F(0) = 0 = F(d), \tag{58}$$
$$f(0) = 0 = f(d).$$

3a. Miles' Theorem

Miles (1961) obtained the theorem that if

$$g\beta > \frac{1}{4} U'^2 \quad \text{or} \quad J(y) \equiv \frac{g\beta}{U'^2} > \frac{1}{4} \tag{59}$$

everywhere in the flow, then the flow is stable, or $c_i = 0$. Here J is called the Richardson number. The following elegant proof of Miles' theorem was given by Howard (1961).

Writing

$$W = U - c \quad \text{and} \quad G = W^{1/2}F,$$

one transforms (55) into

$$(\bar{\rho}WG')' - [\tfrac{1}{2}(\bar{\rho}U')' + k^2\bar{\rho}W + \bar{\rho}W^{-1}(\tfrac{1}{4}U'^2 - g\beta)]G = 0. \tag{60}$$

If the boundaries are fixed,

$$G(0) = 0 = G(d).$$

Multiplying (60) by G^*, integrating, and using the boundary conditions, Howard obtained

$$\int_0^d \left[\bar{\rho} W(|G'|^2 + k^2|G|^2) + \frac{1}{2}(\rho U')'|G|^2 + \bar{\rho}\left(\frac{1}{4}U'^2 - g\beta\right)\left|\frac{G}{W}\right|^2 \right] dy = 0.$$

(61)

If $c_i \neq 0$, the imaginary part of (61) is

$$\int_0^d \bar{\rho}(|G'|^2 + k^2|G|^2)\, dy + \int_0^d \bar{\rho}\left(g\beta - \frac{1}{4}U'^2\right)\left|\frac{G}{W}\right|^2 dy = 0,$$

from which the truth of Miles' theorem is evident. Realizing that $|W|^{-2} \leq c_i^2$, Howard also obtained from the preceding equation the result

$$k^2 \int_0^d \bar{\rho}|G|^2\, dy < \frac{1}{c_i^2} \max\left(\frac{1}{4}U'^2 - g\beta\right) \int_0^d \bar{\rho}|G|^2\, dy$$

or

$$k^2 c_i^2 < \max\left(\frac{1}{4}U'^2 - g\beta\right),$$

(62)

if $c_i \neq 0$. Inequality (62) gives an upper bound for c_i^2. It also contains Miles' theorem.

3b. Howard's Semicircle Theorem

Working with (55) and (58), Howard (1961) succeeded also in showing that, if $c_i \neq 0$,

$$[c_r - \tfrac{1}{2}(a+b)]^2 + c_i^2 < \tfrac{1}{4}(b-a)^2,$$

(63)

in which a is the minimum and b the maximum of U. This is Howard's semicircle theorem, which states that if $c_i > 0$, then c must lie within the semicircle in the upper half of the complex-c plane with the range of U as diameter.

We note that if $\bar{\rho}$ has discontinuities, including the one at the free surface if the upper surface is free, Miles' theorem and Howard's semicircle theorem remain true, provided U remains continuous. All one needs to do is to consider $\bar{\rho}'$ and $[\bar{\rho}(U-c)^2 F']'$ as generalized functions and regard the integrals in (61) and other equations as integrals in the Stieltjes sense.

3c. Sufficient Conditions for Stability

In terms of the dimensionless variables

$$f = \frac{f}{Vd}, \quad \hat{\rho} = \frac{\bar{\rho}}{\rho_0}, \quad \hat{y} = \frac{y}{d}, \quad \hat{U} = \frac{U}{V}, \quad \hat{c} = \frac{c}{V}, \quad \alpha = kd,$$

in which V is a reference velocity and ρ_0 a reference density, equation (57) becomes, after the circumflexes are dropped,

$$(\bar{\rho}f')' + \left(\frac{(\bar{\rho}U')'}{c - U} - \alpha^2 \bar{\rho} - \frac{N\bar{\rho}'}{(c - U)^2} \right) f = 0, \tag{64}$$

in which the primes indicate differentiations with respect to the dimensionless y, and $N = gd/V^2$. We now specify that $\bar{\rho}$ is continuous and the upper surface is fixed. Hence the boundary conditions are

$$f(0) = 0 = f(1). \tag{65}$$

The system consisting of (64) and (65) gives rise to the (complex) secular equation

$$F_1(\alpha, N, c) = 0. \tag{66}$$

When c_i is set to zero and c_r is eliminated from the two real equations contained in (66), we have

$$F_2(\alpha, N) = 0,$$

which gives the neutral stability curve, if one exists. It is possible, however, that c is real for all values of α and N, in which case there is no neutral stability curve or stability boundary in the α-N plane.

In this subsection we shall assume $\bar{\rho}$ and U to be continuous and analytic, and $\bar{\rho}' < 0$ throughout. First, we note that given $\bar{\rho}$, U, and α, the flow is stable for a sufficiently large N according to Miles' theorem, since a large enough N can always be found to make (59) hold. If for some N the flow is unstable, as N is increased there will be one N for which c_i first becomes zero. At this N the c (or c_r) must be within the range of U, since as it is approached through unstable states c_r must be within the range of U according to Howard's semicircle theorem. At the stability boundary (where c first becomes real), (57) is then singular, and we call the mode a singular neutral mode (SNM). The discussion in this paragraph shows that the flow cannot be unstable if singular neutral modes do not exist. To find sufficient conditions for stability, it is then sufficient to find conditions under which singular neutral modes do not exist.

We shall assume U to be monotonic in this subsection. Miles (1961) has shown that singular neutral modes are impossible if the Richardson number $J(y)$ is everywhere greater than $\frac{1}{4}$. In his proof he actually showed that an SNM with a $J(y_c) > \frac{1}{4}$ at $y = y_c$ where $U = c$ is impossible. It can be shown (Yih, 1970) that a SNM with $J(y_c) = \frac{1}{4}$ is also impossible. In that case the two roots of the indicial equation of (57), which has a "regular" singularity at $y = y_c$, are equal, and all that is necessary to construct a proof is to use the solutions, one of which contains a term with the factor $\ln(y - y_c)$, in such an exceptional case. If this latter solution is present, the jump of the Reynolds stress makes the SNM impossible. If it is absent, the proof for $J(y_c) < \frac{1}{4}$ applies. We shall not give the details of the proof here but instead shall concentrate on the case $J(y_c) < \frac{1}{4}$.

For $J(y_c) < \frac{1}{4}$ Miles (1961) gave the solutions of (64) to be

$$f_{\pm}(y) = (y - y_c)^{(1 \pm \kappa)/2} w_{\pm}, \tag{67}$$

in which $w = 1 + A(y - y_c)/(1 \pm \nu) + \cdots$, with

$$\kappa = (1 - 4J_c)^{1/2},$$

$$J_c = J(y_c),$$

$$A = \left[(1 + J) \frac{(\bar{\rho} U')'}{\bar{\rho} U'} - \frac{J\bar{\rho}''}{\bar{\rho}'} + \gamma(\ln \bar{\rho})' \right]_{y_c},$$

$$\gamma = \frac{1 + \kappa}{2}.$$

Miles also showed that for $J_c < \frac{1}{4}$ the solution for an SNM, if one exists, must be either f_+ or f_-. We can demonstrate our point by showing the impossibility of the solution f_+. The demonstration of the impossibility of f_- is similar.

We note first of all that for an SNM which is a stability boundary c (or c_r), being the limit of c_r as $c_i \to 0$, must be bounded away from a or b, since c_r for a nonzero c_i, however small, is bounded away from a or b because (63) is a strict inequality, never an equality. Hence, for our purpose of finding sufficient conditions for stability, it is sufficient to consider the impossibility of an SNM with y_c in $0 < y_c < 1$, with the endpoints zero and 1 ruled out for y_c. The satisfaction of the boundary conditions then demands that $w_\pm$ be zero at $y = 0$ and $y = 1$. The possibility or impossibility of an SNM hinges on the zeros of w_+, which will be denoted by w for brevity. This leads us to consider the differential equation for w, which can be obtained from (64) and (67):

$$(\bar{\rho} z^{2\gamma} w')' + z^{2\gamma} \left[- J_c \bar{\rho} z^{-2} + \gamma \bar{\rho}' z^{-1} + \frac{(\bar{\rho} U')'}{c - U} - \alpha^2 \bar{\rho} - \frac{N \bar{\rho}'}{(c - U)^2} \right] w = 0,$$

$$(68)$$

in which $z = y - y_c$, $\gamma = (1 + \kappa)/2$, and

$$J_c = \left(- \frac{\bar{\rho}'}{\bar{\rho} U'^2} \right)_c N, \tag{69}$$

with $\bar{\rho}$ and U now dimensionless.

We define Q by

$$Q = - \frac{N \bar{\rho}'}{(U - c)^2} - \frac{J_c \bar{\rho}}{z^2},$$

and assume U', $(\bar{\rho} U')'$, and $(\ln \bar{\rho})''$ to be positive. Since $\bar{\rho}' < 0$ and U' and $(\rho U')'$ are positive, U'' is positive. Thus $U - c$ is greater than $U_c z$ for $z > 0$. On the other hand $- \bar{\rho}'/\bar{\rho}$ is less than $(- \bar{\rho}'/\bar{\rho})_c$ for $z > 0$, since $(\ln \bar{\rho})'' > 0$. Thus Q is negative above any point with $y_p > y_c$ if it is negative at $y = y_p$. But for small z

$$Q = \frac{\bar{\rho}_c J_c}{z^2} \left\{ \left[\frac{\bar{\rho}}{\bar{\rho}'} (\ln \bar{\rho})'' - \frac{U''}{U'} \right]_{y_c} z + \cdots \right\}, \tag{70}$$

which is negative for small positive z, hence Q is and, a fortiori, the bracket in (68) is negative above y_c. Integrating (68) between y_c and 1, we have

$$(\bar{\rho} z^{2\gamma} w w')_1 - \int_{yc}^{1} z^{2\gamma} (\bar{\rho} w'^{2\gamma} + G w^2) \, dy = 0, \tag{71}$$

in which $- G$ is the bracket in (69), so that G is positive in the interval of integration. Equation (71) shows clearly that $w(1)$ cannot vanish. Note that the integral in (71) is convergent in spite of the simple pole at $z = 0$ in G, one of which is in Q, as can be seen from (70). Hence we have (Yih, 1970):

THEOREM 1. *If $\bar{\rho}$ and U are continuous and analytic, $\bar{\rho}' < 0$, $U' > 0$, and $(\bar{\rho}U')'$ and $(\ln \bar{\rho})''$ are positive throughout, then singular neutral modes are impossible, and the flow is stable.*

In a similar way, we can prove

THEOREM 2. *If $\bar{\rho}$ and U are continuous and analytic, $\bar{\rho}' < 0$ and $U' < 0$, and U'' and $(\ln \bar{\rho})''$ are negative throughout, then singular neutral modes are impossible, and the flow is stable.*

To prove Theorem 2, it is necessary only to define z as $y_c - y$ and integrate the equation for w corresponding to (68) from zero to y_c.

Theorems 1 and 2 constitute the natural generalization of the well-known Rayleigh theorem for homogeneous fluids.

It remains to note that normal modes may not be adequate for studying stability. But nobody has yet shown a case of instability not described by normal modes, except the obviously unstable case when $\bar{\rho}$ increases upward for a horizontal flow.

3d. Sufficient Conditions for Instability

Let B be the coefficient of f in (64), m be the minimum and M the maximum of $B/\bar{\rho}$ between y_1 and y_2, with $0 < y_1 < y_2 < 1$. For convenience of exposition, we state three conditions, or assumptions, some or all of which will be met or made:

(a) $\bar{\rho}$ and U are continuous and analytic, $\bar{\rho}' \leq 0$, and at any point where $\bar{\rho}'$ vanishes U'' also vanishes, and $U = U_c = $ constant,

(b) $m \geq \dfrac{(n\pi)^2}{(y_2 - y_1)^2}$, $n = $ an integer,

(c) $M \leq \dfrac{(n+1)^2\pi^2}{(y_2 - y_1)^2}$, $n = $ an integer.

Note that if assumption (a) is satisfied, then $\bar{\rho}''$ must vanish where $\bar{\rho}'$ does, for $\bar{\rho}$ cannot increase with z since $\bar{\rho}' \leq 0$. Furthermore, with (a) satisfied, (64) is not singular if $c = U_c$, and we can apply the Sturm-Liouville theory. Applying that theory, we readily obtain two theorems (Yih, 1970):

THEOREM 3. *If assumptions* (a) *and* (b) *are satisfied, there are at least n modes with* $c = U_c$ *and* $\alpha = \alpha_i$ $(i = 1, 2, \ldots, n)$, *and with* α_i *increasing with i. For the i-th mode there are at least* $n - i$ *internal zeros of the eigenfunction f.*

THEOREM 4. *If assumptions* (a), (b), *and* (c) *are satisfied, there are exactly n modes with* $c = U_c$ *and* $\alpha = \alpha_i$ $(i = 1, 2, \ldots, n)$, *and with* α_i *increasing with i. For the i-th mode there are exactly* $n - i$ *internal zeros of the eigenfunction f.*

It can be shown, by an argument similar to that used by Lin (1955), that if we vary α^2 slightly from any of the $\alpha_1, \alpha_2, \ldots, \alpha_n$ mentioned in Theorems 3 and 4, c will become complex. We then have

THEOREM 5. *Near the neutral modes stated in Theorem 3, there are contiguous unstable modes. That is to say, if conditions* (a) *and* (b) *are satisfied, the flow is unstable.*

Incidentally, Theorems 3 and 4 explain why the stability boundary in the α-N plane may be multivalued in α for the same N, as found by Miles (1963).

3e. Nonsingular Modes

It remains to study waves with a real c outside of the range of U, whose minimum and maximum will continue to be denoted by a and b. We now replace assumption (a) with

(d) $\bar{\rho}$ and U are continuous and analytic, $\bar{\rho}' \leq 0$,

and we recall that m and M defined in Subsection 3d depend on c. Since for c outside of the range of U (64) is nonsingular, we can again apply the Sturm-Liouville theory. Doing so, we obtain the following two theorems (Yih, 1970):

THEOREM 6. *If* (d) *and* (b) *are satisfied for* $c = a - \epsilon$, *then there are at least* n *nonsingular modes with* $c = a - \epsilon$, $\alpha = \alpha_i$ $(i = 1, 2, \ldots, n)$, *and* α_i *increasing with* i. *For the* n-*th mode there are at least* $n - i$ *internal zeros for* f. *If* (c) *holds in addition, there are exactly* n *such modes, the* i-*th of which has exactly* $n - i$ *interval zeros of* f. *If* $(\bar{\rho}U')'$ *is negative, then* n *can only increase as the arbitrary constant* ϵ *decreases.*

THEOREM 7. *Theorem 6 remains true if we replace* $c = a - \epsilon$ *with* $c = b + \epsilon$, *and the word* "*negative*" *with the word* "*positive.*"*

* This theorem, as given in Yih (1970), contains a misprint; the word "negative" there should be changed to "positive."

CHAPTER XI

Invariant Functionals of Nonlinear Equations of Evolution

Peter D. Lax

We begin by briefly outlining those aspects of the spectral theory of operators in Hilbert space that are used in Section 1 of this chapter.

A real *Hilbert space*, or *inner product space*, is a linear space over the real numbers in which an inner product (f, g) is defined for every pair of vectors in the space. The properties of inner products are:

(i) *linearity*: $(af + bg, h) = a(f, h) + b(g, h)$, where a, b are scalars, f, g, and h are vectors.

(ii) *symmetry*: $(f, g) = (g, f)$.

(iii) *positivity*: for $f \neq 0$, $(f, f) > 0$.

The quantity $(f, f)^{1/2}$, denoted as $||f||$, is called the *norm* of f. A real-valued function $l(f)$ defined on the space is called a *linear functional* if

$$l(af + bg) = al(f) + bl(g).$$

By property (i) above

$$l(f) = (f, g)$$

is a linear functional of f for every choice of g. We assume that, conversely, every continuous linear functional is of that form; this property of a Hilbert space is called *completeness*.

Prototypes of a Hilbert space are spaces of square integrable functions on an interval, where the scalar product of two functions f and g is defined as

$$(f, g) = \int f(x)g(x) \, dx.$$

This chapter is an expanded version of a paper of the same title which appeared in *Proceedings of the International Conference on Functional Analysis and Related Topics*, 1969. Permission to reprint portions of that paper has been granted by the University of Tokyo Press.

A *linear operator L* is a mapping of the Hilbert space into itself, satisfying

$$L(af + bg) = aLf + bLg.$$

Differential operators play an important role in mathematical physics; these operators are not defined for every function f but only for those that are smooth.

The operator L^* is called *adjoint to L* if

$$(Lf, g) = (f, L^*g)$$

for all f and g for which the operators are defined and if L^* cannot be extended without violating this property. (The notation should not be confused with that for complex conjugation used elsewhere in this book.) An operator is called *self-adjoint* if it is adjoint to itself, *antiself-adjoint* if $-L$ is adjoint to L. If U and U^{-1} are everywhere defined and adjoint to each other, U is called *unitary*. These classes of operators occupy the center of the stage in mathematical physics.

Let U be unitary; then for any vector g,

$$||Ug||^2 = (Ug, Ug) = (g, U^*Ug) = (g, g) = ||g||^2.$$

Thus unitary operators are *length preserving*. The mapping $g = Uh$ can be regarded as merely the introduction of new coordinates (representation) in Hilbert space. From the point of view of this new representation an operator L appears as $U^{-1}LU = L_1$. Observe that if L is self-adjoint, so is L_1. The operators L and L_1 are called *unitarily equivalent*.

The *spectrum* of an operator L is the set of those numbers λ for which $L - \lambda$ is not invertible. If $L - \lambda$ annihilates a nonzero vector n, λ is called an *eigenvalue* and n is called an *eigenvector*. The totality of all eigenvalues is called the *point spectrum* of L_1. The spectrum and point spectrum are often the most eagerly sought information about an operator; in many situations they are susceptible to direct physical observation.

In Section 1 we shall study the classical Schrödinger operator

$$L = D^2 + u;$$

the function u is called the *potential*. If the functions on which L acts lie on a finite interval, boundary conditions have to be imposed at both ends to make the operator L self-adjoint. A classical result asserts that the operator L, acting on a finite interval and subject to such boundary conditions, has a *complete set of eigenvectors, n_j*, where completeness means that the n_j can be introduced as a system of coordinates in Hilbert space.

The point spectrum $\{\lambda_j\}$ is completely determined by the potential u appearing in the operator L. It is well known that the converse is false—that is, that two operators, L and L_1, with entirely different potentials u and u_1 can have the same point spectrum. In the first part of Section 1 we investigate the class of all potentials u_1 for which L_1 has the same spectrum as a given second-order operator L. The main tool used is the observation that *L and L_1 have the same spectrum if and only if they are unitarily equivalent*. We also employ operators which depend on a parameter t and which can be differentiated with respect to t. The derivative of an operator-valued function is defined the same way as the ordinary derivative—that is, as the limit of difference quotients. The ordinary rules of calculus apply, in particular Leibnitz' rule concerning the differentiation of products.

To illustrate this calculus of operator-valued functions, consider a one-parameter family of operators $U(t)$ depending differentiably on t. Unitarity implies that

$$UU^* = I,$$

where I denotes the identity. Differentiating with respect to t and denoting t derivatives by a subscript, we have

$$U_t U^* + UU^*_t = 0.$$

Abbreviating $U_t U^*$ as B, we now make use of these further simple rules:
- (i) $(U^*)^* = U$; in other words, adjointness is reflexive for operators defined everywhere.
- (ii) $(U_t)^* = U^*_t$; in other words, the operations of differentiating with respect to t and taking the adjoint *commute*.
- (iii) $(PQ)^* = Q^*P^*$; in other words, the adjoint of a product of operators is the product of their adjoints *in the reverse order*.

Applying these rules to $B = U_t U^*$, we get

$$B^* = (U_t U^*)^* = (U^*)^*(U_t)^* = UU_t^*.$$

Because U is unitary, we find

$$B + B^* = 0,$$

so that B is *antisymmetric*. This relation will be used in Section 1.

As emphasized in (iii), operator multiplication is not commutative; the order of factors in a product matters. Rather than viewing this with alarm, we point to it with pride; the difference of PQ and QP is called the *commutator* of P and Q and is denoted by the special symbol $[P, Q]$.

In the second part of Section 1 we study the Schrödinger operator over the whole real axis for potentials that tend to zero as $|x| \to \infty$. In this case there is only a finite number of eigenvalues (possibly none), all of which are positive. Every negative number $-k^2$ belongs to the spectrum, but the associated eigenfunctions—that is, solutions of

$$(L + k^2)\Psi = 0,$$

—are not square integrable, so they are not part of Hilbert space. In spite of this defect, they play an important role for the operator L, which is exploited at the end of Section 1.

1. Nonlinear Equations Associated with Linear Operators

It is well known that the spectrum of the Sturm-Liouville operator

$$L = D^2 + u, \qquad \left(D = \frac{d}{dx} \right) \tag{1}$$

does not determine the potential u uniquely—that is, that given any u there are many other potentials u' such that $L' = D^2 + u'$ is unitarily equivalent to L. We start with the infinitesimal version of this problem: find one-parameter families of potentials $u(t)$ such that all members of the one-parameter family of operators $L(t) = D^2 + u(t)$ are unitarily equivalent to each other. This means that there is a one-parameter family of unitary operators $U(t)$ such that

$$U^*(t)L(t)U(t) \tag{2}$$

is independent of t. Here U^* is the adjoint of U. Differentiating (2) with respect to t, we get

$$U_t^*LU + U^*L_tU + U^*LU_t = 0. \tag{3}$$

A one-parameter family of unitary operators satisfies a differential equation of the form

$$U_t = BU,$$ (4)

where $B = B(t)$ is antiself-adjoint:

$$B^* = -B.$$ (5)

Taking the adjoint of (4), we get

$$U_t^* = U^* B^* = -U^* B.$$ (6)

Substituting (4) and (6) into (3), we get, after multiplication by U on the left and U^* on the right,

$$-BL + L_t + LB = 0$$

—that is,

$$L_t = BL - LB = [B, L].$$ (7)

In our case, where we either impose periodic boundary conditions on an interval or take the functions on all of R, in which case no boundary conditions are needed, we have

$$L_t = u_t.$$ (8)

So to solve (7) we have to look for antisymmetric operators B whose commutator with L [defined by (7)] is multiplication. Such an operator is

$$B = B_0 = D.$$

Under periodic boundary conditions, D and D^2 commute, so

$$[B_0, L] = [D, u] = u_x.$$

Equation (7) becomes

$$u_t = u_x,$$ (9)

and we conclude from our general result that if u changes subject to (9), the unitary class of L given by (1) does not change. This is a trivial result, since solutions of (9) are of the form $u = f(x + t)$. Changing the potential subject to (9) merely translates it, which under periodic boundary conditions clearly leads to an equivalent operator.

We get a nontrivial result with

$$B_1 = D^3 + bD + Db,$$

where b is a function to be chosen. A brief calculation gives

$$[B_1, L] = 3u_x D^2 + 3u_{xx} D + u_{xxx} - 4b_x D^2 - 4b_{xx} D - b_{xxx} + 2bu_x.$$

Clearly, the choice of

$$b = \tfrac{3}{4}u$$

makes the coefficients of D^2 and D equal zero, so that $[B_1, L]$ is the multiplication operator. We conclude from our general theorem that if u changes subject to

$$u_t = \tfrac{1}{4}u_{xxx} + \tfrac{3}{2}uu_x, \tag{10}$$

then the operators $L(t)$ are unitarily equivalent.

We can generalize this process and choose for B any operator B_q of form

$$B_q = D^{2q+1} + \sum_1^q (b_j D^{2j-1} + D^{2j-1} b_j).$$

Note that B_q is automatically antisymmetric; this makes $[B_q, L]$ automatically symmetric.

$[B_q, L]$ is a differential operator of degree $2q$: requiring it to be of degree 0 imposes q conditions that uniquely determine the q functions b_j as the result of nonlinear differential operators acting on u. The zero-order term itself is of the form $K_q(u)$, K_q, a nonlinear operator. So we conclude that if u changes subject to

$$u_t = K_q(u), \tag{11}$$

the operators $L(t)$ given by (1) are unitarily equivalent.

The process described here has a fairly general scope. For example, if we replace the scalar operator (1) by a matrix second-order operator where the potential u is a symmetric $n \times n$ matrix, we can obtain a matrix analog of equation (11). Generalizing in another direction, if we take for L the fourth-order operator

$$L = D^4 + DuD + v, \tag{12}$$

then choosing B as

$$B = D^3 + bD + Db,$$

with

$$b = \tfrac{3}{8}u,$$

makes $[B, L]$ a second-order operator of the form

$$[B, L] = DpD + q,$$

where

$$p = P(u, v) = -\tfrac{5}{4}D^4u - \tfrac{3}{4}uDu + 3Dv,$$
$$q = Q(u, v) = D^3v + \tfrac{3}{8}D(uv) - \tfrac{3}{8}uD^3u - \tfrac{3}{8}DuD^2u - \tfrac{3}{8}D^5u.$$

Our general theorem, therefore, shows that if u and v change subject to

$$u_t = P(u, v), \qquad v_t = Q(u, v),$$

the operators $L(t)$ given by (12) are unitarily equivalent.

So far this method has not been extended to operators L in more space dimensions.

We return now to the operator L given by (1); its eigenvalues λ_n are *functionals* $\lambda_n[u]$ of the potential L. Since unitarily equivalent operators have the same eigenvalues, it follows that the eigenvalues of (8),

$$\lambda_n[u], \tag{13}$$

are *invariant functionals, called integrals, of the differential equations* (11).

For the simplest and most important of these equations (10), this has been discovered in an entirely different way by Gardner, Greene, Kruskal, and Miura (1967) and is described in Chapter VIII.

There are several reasons why one is interested in finding all the integrals of an equation of evolution. First of all, these integrals are conserved quantities and therefore might very well have a direct physical significance. Total mass, momentum, angular momentum, energy, and charge are splendid examples of physically significant conserved quantities.

Second, conserved quantities are an important mathematical tool; they give a priori estimates that are at the heart of every existence theorem (except for those cases where the solution is explicitly written). For example, the existence theory for hyperbolic equations can be based exclusively on the law of conservation of energy. It is reasonable, therefore, to try to derive a priori estimates from other integrals.

Next we show for the example (10) that knowing the value of the integrals (13) gives interesting information about the behavior of the solution u for $|t|$ large. We introduce a simple change of scale in u and x to bring (10) into the more familiar form

$$u_t + uu_x + u_{xxx} = 0, \tag{14}$$

introduced by Korteweg and deVries for the description of surface waves in water (See Chapter IV). It is well known that this equation has solitary-wave solutions—that is, solutions of the form $u(x, t) = s(x - ct)$, where $s(-\infty) = s(\infty) = 0$. Such a wave exists for every positive speed and is unique except for an arbitrary translation. We normalize s by choosing $x = 0$ to be the point of symmetry of $s = s(x; c)$.

Zabusky and Kruskal (1965) made the remarkable discovery that all solutions of the Korteweg-deVries (K-dV) equation defined on the whole axis and vanishing at $x = \pm\infty$ contain solitary waves, in this sense. To any such solution there corresponds a finite number of speeds $c_1, \ldots, c_N$, which may be called the eigenspeeds of u, such that

$$\lim_{t \to \pm\infty} u(x - c_j t, t) = s(x - \theta^\pm, c_j), \qquad j = 1, \ldots, N, \tag{15}$$

uniformly on bounded sets of x. The meaning of (15) is that for large $|t|$ and around the point $x = c_j t$, u looks like a solitary wave traveling with speed c_j, shifted by $\theta_j^\pm$.

The eigenspeeds c_j are functionals of the solution; clearly, $c_j[u]$ is an invariant functional.

Gardner, Greene Kruskal, and Miura (1967) showed that these invariant functionals are essentially the same as those found before in (13):

$$4\lambda_j[u] = c_j[u].$$

We shall present a simple proof of one part of this proposition (Lax, 1968); that is, if c_j satisfies (15), then $\lambda_j = c_j/4$ is an eigenvalue of the operator $D^2 + u/6$ [the factor is necessitated by the rescaling of equation (10)].

First we show that $\lambda[s(c)] = c/4$. This can be done by explicitly verifying that

$$L[s]w = \frac{c}{4}w$$

is satisfied by $w(x) = \sqrt{s(x; c)}$. Suppose now that (15) holds; then for t large, $u(x) \cong s(x - ct - \theta; c)$ on a long interval around $x = ct$. Therefore,

$$L[u(t)]w \cong \frac{c}{4}w \tag{16}$$

is approximately satisfied on that interval by

$$w(x) = \sqrt{s(x - ct - \theta; c)}.$$

Outside of that interval both sides of (16) are small and tend to zero rapidly. Therefore,

$$||L[u(t)]w - \frac{c}{4}w|| < \epsilon(t)||w||, \tag{17}$$

where $\epsilon(t) \to 0$ as $t \to \infty$. It follows by spectral theory from (17) that a point λ exists in the spectrum of $L[u(t)]$ such that

$$\left| \lambda - \frac{c}{4} \right| < \epsilon(t).$$

Since we have already shown that the spectrum of $L[u(t)]$ is independent of t, it follows that $c/4$ is in the spectrum of L and, being a positive quantity, is the point spectrum of L.

In this proof few special properties of (14) were used. Another series of invariant functionals for (14) can be constructed from an infinite series of conservation laws that can be deduced from (14). These are described in Miura, Gardner, and Kruskal (1968).

The change of scale that turns (10) into (14) and the operator (1) into

$$L = D^2 + \frac{u}{6} \tag{18}$$

turns B into

$$B = -4D^3 - uD - \tfrac{1}{2}u_x. \tag{19}$$

The present derivation can be used to determine, more directly than in Chapter VIII, how the eigenfunctions of $L(t)$ vary with t. Denote by ϕ_0 an eigenfunction of $L(0)$:

$$L(0)\phi_0 = \lambda\phi_0.$$

Since by construction of U, (2) is independent of t, we have [assuming that $U(0) = I$]

$$U^*(t)L(t)U(t)\phi_0 = L(0)\phi_0 = \lambda\phi_0$$

for all t. Multiplying by $U(t)$ and introducing

$$\phi(t) = U(t)\phi_0, \tag{20}$$

we deduce that

$$L(t)\phi(t) = \lambda\phi(t),$$

which shows that the eigenfunctions $\phi(t)$ of $L(t)$ are related to those of $L(0)$ by (20). Differentiating (20) with respect to t and using the differential equation (4) satisfied by $U(t)$, we deduce that

$$\phi_t = B\phi.$$

For the case at hand when B is given by (19), it follows that ϕ satisfies

$$\phi_t + 4\phi_{xxx} + u\phi_x + \tfrac{1}{2}u_x\phi = 0. \tag{21}$$

Now ϕ itself satisfies the eigenvalue equation

$$\phi_{xx} + \tfrac{1}{6}u\phi = \lambda\phi.$$

Differentiating this with respect to x, multiplying by 4, and subtracting from (21) gives

$$\phi_t + (4\lambda + \tfrac{1}{3}u)\phi_x - \tfrac{1}{6}u_x\phi = 0, \tag{22}$$

a first-order equation satisfied by ϕ.

On the infinite interval $(-\infty, \infty)$ and for a function u which tends to 0 rapidly as $x \to \pm\infty$, the operator L given by (18) has a finite number of positive eigenvalues and a continuous spectrum of multiplicity covering the negative real numbers. With each point $-k^2$ of the continuous spectrum we can associate two *improper eigenfunctions*, each a solution of

$$\phi_{xx} + \tfrac{1}{6}u\phi + k^2\phi = 0. \tag{23}$$

We saw earlier that the proper eigenfunctions ϕ of $L(t)$ satisfy the differential equation (22). It is not difficult to show the converse:

LEMMA 1. *Let ϕ be a solution of*

$$\phi_t + (\tfrac{1}{3}u - 4k^2)\phi_x - \tfrac{1}{6}u_x\phi = 0, \tag{24}$$

which at $t = 0$ satisfies

$$[L(0) + k^2]\phi(0) = 0.$$

Then ϕ satisfies

$$[L(t) + k^2]\phi(t) = 0$$

for all t.

Proof. Define χ by

$$\chi = (L + k^2)\phi.$$

Equation (24) can be rewritten as

$$\phi_t - B\phi - 4D\chi = 0, \tag{25}$$

where B is defined in (19). We now calculate χ_t. Using (25), we get

$$\chi_t = L_t\phi + L\phi_t + k^2\phi_t$$
$$= L_t\phi + LB\phi + 4LD\chi + k^2B\phi + 4k^2D\chi.$$

Using the identity (7), according to which $L_t = BL - LB$, we get, after grouping terms,

$$\chi_t = B(L + k^2)\phi + 4(L + k^2)D\chi$$
$$= B\chi + 4(L + k^2)D\chi.$$

This equation can be rewritten as

$$\chi_t + (\tfrac{1}{3}u - 4k^2)\chi_x + \tfrac{1}{2}u_x\chi = 0. \tag{26}$$

Since by assumption $\chi(0) = 0$, and since solutions of (26) are uniquely determined by their initial values, it follows that $\chi(t) = 0$ for all t, as asserted in Lemma 1.

It is well known that if u tends to zero fast enough as $|x| \to \infty$, then solutions of (23) behave like linear combinations of e^{ikx} and e^{-ikx}. Suppose we normalize $\phi(0)$ so that it consists of a plane wave of unit strength coming from, say, the left, scattered by the potential $u(0)$:

$$\phi(x, 0, k) \simeq \begin{cases} e^{ikx} + Re^{-ikx} & \text{for } x \text{ near } -\infty, \\ Te^{ikx} & \text{for } x \text{ near } +\infty. \end{cases} \tag{27}$$

The quantity $R = R(k)$ is called the *reflection coefficient* and $T = T(k)$ the *transmission coefficient*.

How does the asymptotic behavior of ϕ near $|x| = \infty$ vary with time? For $|x|$ large we can write

$$\phi(x, t) \simeq A(t)e^{ikx} + B(t)e^{-ikx},$$

and we can neglect u and u_x in (24). We get

$$(A_t - 4k^3 iA)e^{ikx} + (B_t + k^3 iB)e^{-ikx} = 0,$$

which implies that

$$A(t) = e^{4ik^3 t}, \qquad B(t) = e^{-4ik^3 t}.$$

Taking the initial values (27) of ϕ into account, we conclude that

$$\Psi = e^{-4ik^3 t}\phi$$

represents, for each t, a plane wave of unit strength coming from the left and scattered by the potential $\frac{1}{6}u(t)$:

$$\Psi(x, t, k) = \begin{cases} e^{ikx} + e^{-8ik^3 t}R(k)e^{-ikx} & \text{for } x \text{ near } -\infty, \\ T(k)e^{ikx} & \text{for } x \text{ near } +\infty. \end{cases}$$

This asymptotic description of Ψ shows that *the transmission coefficient of $L(t)$ is independent of t, and the reflection coefficient R varies exponentially in t*:

$$R(k, t) = R(k, 0)e^{-8ik^3 t}. \tag{28}$$

Further details may be found in Lax (1970). This result is originally due to Gardner et al. (1967). Chapter VIII points out that the following method, based on relation (28), can be used to solve the initial-value problem for the Korteweg-deVries equation (14).

Given the initial value $u(x, 0)$, one can determine the point eigenvalues as well as the transmission and reflection coefficients of the operator $L(0)$. Using the previous results, we deduce that $L(t)$ has the same point eigenvalues as $L(0)$, and its reflection coefficient is given by (28). Using the Gel'fand-Levitan procedure for solving the inverse problem, we can reconstruct the potential $\frac{1}{6}u(x, t)$ appearing in the operator $L(t)$.

Suppose u is subject to a linear equation of evolution:

$$u_t = Au.$$

Let ϕ be an eigenfunction of A^*:

$$A^*\phi = \lambda\phi.$$

The linear functional $r(t) = [u(t), \phi]$ then depends exponentially on t:

$$r(t) = e^{\lambda t}r(0).$$

The analysis above exhibits a family of nonlinear functionals $R[u]$ that vary exponentially in t if u is subject to the Korteweg-deVries equation.

2. Nonlinear Hyperbolic Systems of Conservation Laws

A conservation law is an equation of the form

$$u_t + f_x = 0. \tag{29}$$

Here we shall take f to be some nonlinear function of u: $f = f(u)$. Denoting

$$\frac{df}{du} = a(u), \tag{30}$$

we can write (29) as

$$u_t + a(u)u_x = 0, \tag{31}$$

which asserts that u is constant along trajectories that propagate with speed a. For this reason a is called the *signal speed*. Note that, owing to the assumed nonlinearity, the signal speed depends on u.

We shall consider weak solutions of (29), as described in Chapter III. For bounded piecewise continuous solutions, this can be expressed in the integral form

$$\int_{-\infty}^{y} u(x, t_2)\, dx = \int_{-\infty}^{y} u(x, t_1)\, dx - \int_{t_1}^{t_2} f[u(y, t)]dt. \tag{32}$$

Here we have assumed that $u(-\infty, t) = 0$, and that

$$f(0) = 0, \tag{33}$$

which can always be achieved by subtracting a constant from f. The physical interpretation of (32) is that the change in the amount of u contained in $(-\infty, y)$ during the time interval (t_1, t_2) is due to the *flux* across the endpoint y.

For piecewise continuous solutions, (32) implies the *jump condition*

$$s[u]_j = [f]_j \tag{34}$$

where $[u]_j = u_l - u_r$, $[f]_j = f_l - f_r$, the subscripts *l* and *r* indicating values taken on the left and right edges of the discontinuity, and *s* denotes the speed of propagation of the line of discontinuity $x = x(t)$—that is,

$$s = \frac{dx}{dt}.$$

To insure a unique solution we impose the *entropy condition*, which for piecewise continuous solutions demands that signals originating near a discontinuity always reach the point of discontinuity. As shown in Chapter III, this is equivalent to

$$a(u_l) > s > a(u_r).$$

Using the definition (30) of *a* and (34) of *s* means that

$$f_u(u_l) > \frac{f(u_l) - f(u_r)}{u_l - u_r} > f_u(u_r).$$

For *f* convex this is equivalent to

$$u_l < u_r. \tag{35}$$

It is well known (see Chapter III) that for *f* convex and for bounded initial data with compact support, the initial-value problem is properly posed within the class of solutions that satisfy the entropy condition. We turn now to the asymptotic behavior of these solutions as $t \to \infty$.

We start with an example. Take the simplest nonlinear function

$$f(u) = cu + \frac{k}{2}u^2,$$

which is convex if $k > 0$. In this case the differential form (31) of the equation is

$$u_t + (c + ku)u_x = 0. \tag{36}$$

The jump relation (34) takes the form

$$s = c + k\frac{u_l + u_r}{2}. \tag{37}$$

It is not difficult to verify that for any values of the parameters p, $q \geq 0$, the function $v = v(x, t; p, q)$ defined by

$$v(x, t) = \begin{cases} 0 & \text{for } x < ct - \sqrt{pt}, \\ \dfrac{1}{k}\left(\dfrac{x}{t} - c\right) & \text{for } ct - \sqrt{pt} < x < ct + \sqrt{qt}, \\ 0 & \text{for } ct + \sqrt{qt} < x, \end{cases} \tag{38}$$

satisfies the differential equation (36) whenever v is continuous and satisfies the jump relation (37) along the discontinuities

$$x_- = ct - \sqrt{pt} \quad \text{and} \quad x_+ = ct + \sqrt{qt}.$$

Expression (38) is the "N-wave" described in Chapter IV, Section 2. Since the parameters p and q as well as k are non-negative, it follows that the entropy condition (35) is satisfied.

The significance of the two-parameter family of solutions v given by (38) is that every solution u of (29) with bounded integrable initial values behaves for large t like some v. More precisely, we have

THEOREM 1. *Suppose f is convex. To every solution u of (29) with bounded integrable initial values there corresponds two non-negative constants p and q such that*

$$\lim_{t \to \infty} ||u(t) - v(t; p, q)||_1 = 0,$$

where the norm is the L_1 norm in x.

For (36) this was proved by Hopf (1950), and for any strictly convex f by Lax (1957). For arbitrary convex f the constants c and k have to be taken as

$$c = f'(0), \qquad k = f''(0).$$

It is easy to verify that for fixed h

$$\lim_{t \to \infty} ||v(t + h; p, q) - v(t)||_1 = 0.$$

This shows that the parameters p and q entering the asymptotic description of u are invariant functionals of u. We proceed now to determine them.

We remark that it follows from the conservation law (32) that for solutions u that are zero near $x = \pm\infty$,

$$I_0[u] = \int_{-\infty}^{\infty} u(x, t)\, dx \tag{39}$$

is an invariant functional. It follows from Theorem 1 that

$$I_0(u) = I_0[v(p, q)].$$

A short calculation gives

$$I_0(v) = \frac{q - p}{k};$$

therefore,

$$q - p = kI_0(u).$$

To determine q and p separately we need another invariant functional. Such a one is given in

THEOREM 2. *For f convex*

$$I_1[u] = \min_y \int_{-\infty}^{y} u(x, t)\, dx \tag{40}$$

is an invariant functional for solutions of (29).

A proof of this is given in Lax (1957). We present here another derivation, valid for a piecewise continuous function whose initial data have compact support. Denote by $\tilde{M}(t)$ the value of the right side (40), and by $y(t)$ any value where the minimum is taken on. Taking y in (32) to be $y(t_1)$, and $y(t_2)$, we obtain, using the definition of $\tilde{M}(t)$ as a minimum, the pair of inequalities

$$\tilde{M}(t_2) \le \tilde{M}(t_1) + \int_{t_1}^{t_2} f(y_1)\, dt$$

and

$$\tilde{M}(t_2) \ge \tilde{M}(t_1) + \int_{t_1}^{t_2} f(y_2)\, dt,$$

which implies

$$|\tilde{M}(t_1) - \tilde{M}(t_2)| \le |t_1 - t_2|F, \tag{41}$$

where

$$F = \sup_{t_1 < t < t_2} |f(y_1, t)|,\ |f(y_2, t)|. \tag{42}$$

Here $f(y, t)$ abbreviates $f[u(y, t)]$, and $y_1 = y(t_1)$, $y_2 = y(t_2)$.

LEMMA 2. *The point $y(t)$, t is a point of continuity of u, and*

$$u(y(t), t) = 0. \tag{43}$$

Proof. By the minimizing property of $y(t)$ we must have

$$u_l(y, t) \leq 0 \quad \text{and} \quad u_r(y, t) \geq 0.$$

This contradicts the entropy condition (35) unless $u_l = u_r = u = 0$, as asserted.

Since

$$\int_{-\infty}^{y} u(x, t) \, dx$$

is a continuous function of y and t, independent of x and t for $|x|$ large, it follows that the set of minimizing points $y(t)$, t with $t \leq T$ forms a compact set. From this and Lemma 2 we easily deduce this:

COROLLARY. *The set of minimizing points $y(t)$, t with $t \leq T$ has a positive distance $d(\epsilon)$ from any shock of strength $\geq \epsilon = |u_l - u_r|$.*

We now take t_1, $t_2 \leq T$, so that $|t_1 - t_2| < d(\epsilon)$, and so that over any t interval of length $d(\epsilon)$ that does not cross a shock, the oscillation of u is $< \epsilon$. It follows that the oscillation of $u(y_1, t)$ and of $u(y_2, t)$ over (t_1, t_2) is $< 2\epsilon$. Since by (43) $u(y_1, t_1) = u(y_2, t_2) = 0$, it follows that $|u| < 2\epsilon$ over these t intervals, and since by (33) $f(0) = 0$ we conclude by the continuity of f that $|f| < \eta(\epsilon)$ on these t intervals, where $\eta(\epsilon)$ tends to zero with ϵ. Thus we have from (42) and (41) that

$$|\tilde{M}(t_1) - \tilde{M}(t_2)| < |t_1 - t_2| \eta$$

for $|t_1 - t_2|$ small enough. From this it follows that the derivative of $\tilde{M}$ is zero everywhere, so that $\tilde{M}$ is constant, as asserted in Theorem 2.

A simple calculation gives

$$I_1(v) = -\frac{p}{k}.$$

Therefore, it follows from Theorems 1 and 2 that for any u

$$p = -kI_1(u).$$

Note that the functionals I_0 and I_1 are continuous in the L_1 topology. It follows from Theorem 1 that every other invariant functional that i continuous in the L_1 topology is a function of I_0 and I_1.

We turn now to systems of conservation laws:

$$\frac{\partial}{\partial t}u_i + \frac{\partial}{\partial x}f_i = 0, \qquad i = 1, \ldots, n, \tag{44}$$

with each f_i a nonlinear function of $u_1, \ldots, u_n$. We assume that (44) is hyperbolic system, which means that the matrix

$$A = (a_{ji}) = \left(\frac{\partial f_i}{\partial u_j}\right)$$

has real and distinct eigenvalues for all u. These eigenvalues $c_j = c$ $(u_1, \ldots, u_n)$ are the signal speeds associated with the system (44).

It is reasonable to conjecture that the analog of Theorem 1 holds where v is the sum of n sawtooth functions of the form (38), one fo each of the n signal speeds $c_j(0)$, $j = 1, \ldots, n$. Such a theorem can b proved rigorously for $n = 2$ and for solutions whose initial values ar small enough using the methods developed in Glimm and Lax (1969)

It would follow from the generalization of Theorem 1 to system that the quantities p_j and q_j appearing in the asymptotic description ar invariant functionals. From the conservation form of equations (44 it follows that $I_0[u_j]$ defined by (39) $j = 1, \ldots, n$ are invariant functionals The existence of $2n$ functionals p_j, q_j indicates that the functional defined by (40) also has an analog. Although it is too much to expec an explicit formula, a proof of the existence of such functionals i highly desirable. A more complete account may be found in La (1973).

3. Nonlinear Wave Equations

The nonlinear wave equations to be considered here are of the forr

$$u_{tt} - \Delta u + up(|u|^2) = 0, \tag{45}$$

where $p = p(s)$ is a positive potential. Denote by m the quantity

$$p(0) = m.$$

We shall study complex valued solutions of this equation.

For a certain class of potentials it is known that all solutions tend to zero uniformly in the maximum norm as $|t| \to \infty$. For such solutions the equation satisfied by them, is for large $|t|$, very nearly linear—that is,

$$u_{tt} - \Delta u + mu \cong 0. \tag{46}$$

Morawetz and Strauss (1972) have shown that such solutions tend asymptotically as $t \to \pm\infty$ to a pair of solutions of the linear equation (46). In this section we reproduce a very simple argument of Roffman (1970), which shows that for certain potentials solutions exist that do not tend to zero in the maximum norm as $|t| \to \infty$ and therefore do not behave for large $|t|$ as solutions of the linear equation (46).

THEOREM 3. *Denote by $P(s)$ the integral of $p(s)$:*

$$\frac{dP}{ds} = p(s), \qquad P(0) = 0. \tag{47}$$

Suppose that a value s_0 exists such that

$$P(s_0^2) < ms_0^2. \tag{48}$$

Then (45) has solutions u for which

$$\max_x |u(x, t)| > \delta > 0$$

for all t.

Roffman's proof is based on a relation between two *classical invariant functionals* of (45), *energy E* and *charge Q*, defined as follows:

$$E = \tfrac{1}{2} \int \{ |u_t|^2 + \sum |u_{xj}|^2 + P(|u|^2) \} \, dx$$

and

$$Q = \text{Im} \int u_t \bar{u} \, dx,$$

where here $\bar{u}$ is the complex conjugate of u.

It follows from (47) that if

$$|u| < \delta, \tag{49}$$

then

$$P(|u|^2) > (m - \epsilon)|u|^2, \tag{50}$$

where ϵ tends to zero if δ does. It follows from the Schwarz inequality that

$$Q \leq \frac{\sqrt{m}}{2} \int |u|^2 \, dx + \frac{1}{2\sqrt{m}} \int |u_t|^2 \, dx.$$

Using the definition of E and (50), we conclude that if (50) holds, then

$$Q \leq \frac{1}{\sqrt{m - \epsilon/\sqrt{m}}} E. \tag{51}$$

Consider the following initial data:

$$u(x, 0) = \begin{cases} s_0 & \text{for } |x| < R, \\ s_0|x| - R & \text{for } R < |x| < R + 1, \\ 0 & \text{for } R + 1 < |x|, \end{cases} \tag{52}$$

$$u_t(x, 0) = -iu(x, 0). \tag{53}$$

Clearly, if (48) holds, then for R large enough

$$Q_0 > \frac{1}{\sqrt{m}} E_0, \tag{54}$$

where Q_0 and E_0 are the charge and energy of these initial data. Now (54) contradicts (51) for ϵ small enough. Since (51) holds whenever (49) holds for all x with δ small enough, it follows that the solution u with initial data (52), (53) does not satisfy (49) for δ small enough for any value of t. This proves Theorem 3.

References

Abramowitz, M., and I. A. Stegun. 1965. *Handbook of Mathematical Functions.* New York: Dover. Also available as AMS 55 (1964) from Superintendent of Documents, U.S. Govt. Printing Office, Washington, D.C.

Ball, F. K. 1964. Energy transfer between external and internal gravity waves. *J. Fluid Mech.* 19: 465–78.

Benjamin, T. B. 1966. Internal waves of finite amplitude and permanent form. *J. Fluid Mech.* 25: 241–70.

Benjamin, T. B. 1967. Instability of periodic wave trains in nonlinear dispersive systems. *Proc. Roy. Soc.* A 299: 59–75.

Benjamin, T. B., and J. E. Feir. 1967. The disintegration of wave trains on deep water. *J. Fluid Mech.* 27: 417–30.

Benjamin, T. B., and M. J. Lighthill. 1954. On cnoidal waves and bores. *Proc. Roy. Soc.* A 224: 448–60.

Benney, D. J. 1962. Non-linear gravity wave interactions. *J. Fluid Mech.* 14: 577–84.

Bôcher, M. 1917. *Leçons sur les méthodes de Sturm dans la théorie des équations différentielles linéaires et leurs développements modernes.* Paris: Gauthier-Villars.

Boussinesq, J. 1872. Théorie des ondes et des remous qui se propagent le long d'un canal rectangulaire horizontal, en communiquant au liquide contenu dans ce canal des vitesses sensiblement pareilles de la surface au fond. *J. Math. Pures Appl.* 17 (Series 2): 55–108.

Bretherton, F. P. 1964. Resonant interactions between waves. The case of discrete oscillations. *J. Fluid Mech.* 20: 457–79.

Bretherton, F. P., and C. J. R. Garrett. 1969. Wavetrains in inhomogeneous moving media. *Proc. Roy. Soc.* A 302: 529–54.

Burgers, J. M. 1948. A mathematical model illustrating the theory of turbulence. *Advanc. Appl. Mech.* 1: 171–99.

Carrier, G. F., M. Krook, and C. E. Pearson. 1966. *Functions of a Complex Variable*. New York: McGraw-Hill.

Chikwendu, S. C., and J. Kevorkian. 1972. A perturbation method for hyperbolic equations with small nonlinearities. *SIAM J. Appl. Math.* 22: 235–58.

Cole, J. D. 1951. On a quasi-linear parabolic equation occurring in aerodynamics. *Quart. Appl. Math.* 9: 225–36.

Cole, J. D. 1968. *Perturbation Method in Applied Mathematics*. Waltham, Mass.: Blaisdell.

Conley, C. C., and J. A. Smoller. 1970. Viscosity matrices for two-dimensional nonlinear hyperbolic systems. *Comm. Pure Appl. Math.* 23: 867–84.

Conway, E., and J. A. Smoller. 1966. Global solutions of the Cauchy problem for quasi-linear first-order equations in several space variables. *Comm. Pure Appl. Math.* 19: 95–105.

Copson, E. T. 1965. *Asymptotic Expansions*. Cambridge: Cambridge University Press.

Courant, R., and K. O. Friedrichs. 1948. *Supersonic Flow and Shock Waves*. New York: Interscience.

Courant, R., and D. Hilbert. 1962. *Methods of Mathematical Physics*. Vol. II. New York: Interscience.

Dafermos, C. M. 1972. Polygonal approximations of solutions of the initial value problem for a conservation law. *J. Math. Anal. and Applications*. 38: 33–41.

Drazin, P. G., and D. W. Moore. 1967. Steady two-dimensional flow of fluid of variable density over an obstacle. *J. Fluid Mech.* 28: 353–70.

Dubreil-Jacotin, M. L. 1932. Sur les ondes de type permanent dans les liquides hétérogènes. *Atti Accad. Lincei, Rend. Cl. Sci. Fis. Mat. Nat.* (6) 15: 814–19.

Dubreil-Jacotin, M. L. 1935. Complément à une note antérieure sur les ondes de type permanent dans les liquides hétérogènes. *Atti Accad. Lincei. Rend. Cl. Sci. Fis. Mat. Nat.* (6) 21: 344–46.

Erdélyi, A. 1956. *Asymptotic Expansions*. New York: Dover.

Freund, D. D. 1972. A note on Kaplun limits and double asymptotics. *Proc. Am. Math. Soc.* 35: 464–70.

Fröman, N., and P. O. Fröman. 1965. *JWKB Approximation*. Amsterdam: North Holland Publishing Co.

Gardner, C. S., J. M. Greene, M. D. Kruskal, and R. M. Miura. 1967. Method for solving the Korteweg-deVries equation. *Phys. Rev. Letters* 19: 1095–97.

Gardner, C. S., J. M. Greene, M. D. Kruskal, and R. M. Miura. Korteweg-deVries equation and generalizations. VI. Methods for exact solution. (To be published)

Gardner, C. S., and G. K. Morikawa. 1960. Similarity in the asymptotic behavior of collision-free hydromagnetic waves and water waves. Courant Institute of Mathematical Sciences Report No. NYO-9082. (Unpublished)

Gel'fand, I. M. 1963. Some problems in the theory of quasilinear equations. *Am. Math. Soc. Trans.* (Series 2) 29: 295–381.

Gel'fand, I. M., and S. V. Fomin. 1963. *Calculus of Variations.* Englewood Cliffs, N.J.: Prentice-Hall.

Gel'fand, I. M., and B. M. Levitan. 1955. On the determination of a differential equation from its spectral function. *Amer. Math. Soc. Trans.* (Series 2) 1: 253–304.

Glimm, J. 1965. Solutions in the large for nonlinear hyperbolic systems of equations. *Comm. Pure Appl. Math.* 18: 697–715.

Glimm, J., and P. D. Lax. 1970. Decay of solutions of systems of nonlinear hyperbolic conservation laws. *Memoirs of the Amer. Math. Society.* No. 101. Providence: American Mathematical Society.

Görtler, H. 1943. Über eine Schwingungsersch einung in Flüssigkeiten mit stabiler Dichteschichtung. *Z. angew Math. Mech.* 23: 65–71.

Greenberg, J. M. 1970. On the interaction of shocks and simple waves of the same family. *Arch. Rat. Mech. Analysis* 37: 136–60.

Groen, P. 1948. Contribution to the theory of internal waves. *Mededelingen en Verhandelingen* Serie B. Deel II, No. II. Koninklijk Nederlands Meteorologisch Instituut de Bilt, Nederland.

Hasselmann, K. 1967. A criterion for non-linear wave stability. *J. Fluid Mech.* 30: 737–39.

Havelock, T. H. 1914. *The Propagation of Disturbances in Dispersive Media.* Cambridge: Cambridge University Press. (Republished in 1964 by Stechert-Hafner Service Agency, New York.)

Hayes, W. D. 1970a. Conservation of action and modal wave action. *Proc. Roy. Soc.* A 320: 187–208.

Hayes, W. D. 1970b. Kinematic wave theory. *Proc. Roy. Soc.* A 320: 209–26.

Hayes, W. D. 1973. Group velocity and nonlinear dispersive wave propagation. *Proc. Roy. Soc.* A 332: 199–221.

Hopf, E. 1950. The partial differential equation $u_t + uu_x = \mu u_{xx}$. *Comm. Pur- Appl. Math.* 3: 201–30.

Howard, L. N. 1961. Note on a paper of John W. Miles. *J. Fluid Mech.* 10: 509–12.

Hung, C.-M. and R. Seebass. 1973. Reflexion of a weak shock wave with vibrational relaxation. (To appear)

Hurd, A. E. 1969. A uniqueness theorem for weak solutions of symmetric quasilinear hyperbolic systems. *Pacific J. Math.* 28: 555–59.

Hurd, A. E. 1970. A uniqueness theorem for second-order quasilinear hyperbolic equations. *Pacific J. Math.* 32: 415–27.

Ibbetson, A., and N. Phillips. 1967. Some laboratory experiments on Rossby waves in a rotating annulus. *Tellus* 19: 81–7.

Jeffrey, A., and T. Kakutani. 1972. Weak nonlinear dispersive waves: a discussion centered around the Korteweg-deVries equation. *SIAM Review* 14: 582–643.

Johnson, J. L., and J. A. Smoller. 1967. Global solutions for certain systems of quasi-linear hyperbolic equations. *J. Math. Mech.* 17: 561–76.

Johnson, R. S. 1970. A nonlinear equation incorporating damping and dispersion. *J. Fluid Mech.* 42: 49–60.

Kaplun, S., 1967. *Fluid mechanics and singular perturbations.* New York: Academic Press.

Kaplun, S., and P. A. Lagerstrom. 1957. Asymptotic expansion of Navier-Stokes solutions for small Reynolds number. *J. Math. Mech.* 6: 585–93.

Kay, I. 1955. The inverse scattering problem. New York University, Institute of Mathematical Sciences, Division of Electromagnetic Research Report No. EM-74. (Unpublished)

Kay, I., and H. E. Moses. 1956. The determination of the scattering potential from the spectral measure function. III. Calculation of the scattering potential from the scattering operator for the one-dimensional Schrödinger equation. *Il Nuovo Cimento* (Series X) 3: 276–304.

Kelvin, W. T. 1887. On the waves produced by a single impulse in water of any depth, or in a dispersive medium. *Proc. Roy. Soc.* A 42: 80–85.

Kevorkian, J. 1966. The two variable expansion procedure for the approximate solution of certain nonlinear differential equations. *Lectures in Applied Mathematics* 7, *Space Mathematics III.* Providence, R.I.: American Mathematical Society. Pages 206–75.

Korteweg, D. J., and G. deVries. 1895. On the change of form of long waves advancing in a rectangular canal, and on a new type of long stationary waves. *Phil. Mag.* 39: 422–43.

Kruskal, M. D. 1963. Asymptotology. *Proc. Conf. on Mathematical Models in the Physical Sciences,* University of Notre Dame, South Bend, Ind. 1962. Eds. S. Drobot and P. A. Viebrock, Englewood Cliffs, N.J.: Prentice Hall. Pages 17–48.

Kruskal, M. D. 1965. Asymptotology in numerical computation: Progress and plans on the Fermi-Pasta-Ulam problem. *Proc. IBM Scientific Computing Symposium on Large-Scale Problems in Physics*. Thomas J. Watson Research Center, Yorktown Heights, N.Y. 1963. White Plains: IBM Data Processing Division. Pages 43–67.

Kruskal, M. D., and R. M. Miura. Korteweg-deVries equation and generalizations. VII. Nonexistence of polynomial conservation laws for generalizations. (To be published)

Kruskal, M. D., R. M. Miura, C. S. Gardner, and N. J. Zabusky. 1970. Korteweg-deVries equation and generalizations. V. Uniqueness and nonexistence of polynomial conservation laws. *J. Math. Phys.* 11: 952–60.

Kruskal, M. D., and N. J. Zabusky. 1963. Progress on the Fermi-Pasta-Ulam nonlinear string problem. Princeton Plasma Physics Laboratory Annual Report MATT-Q-21. 301–308. (Unpublished)

Kružkov, S. N. 1970. First order quasilinear equations in several independent variables. *Mat. Sbornik* 81 (123), No. 2. English translation: *Math. USSR Sbornik* 10: 217–43.

Kuznetsov, N. N. 1967. Weak solution of the Cauchy problem for a multidimensional quasi-linear equation. *Mat. Zametki* 2: 401–10. English translation: *Math. Notes* 2: 733–39.

Lamb, H. 1932. *Hydrodynamics*. (6th Ed.) Cambridge: Cambridge University Press. Also, New York: Dover.

Landau, L. D., and E. M. Lifshitz. 1958. *Quantum Mechanics, Non-relativistic Theory*. London: Pergamon Press.

Lax, P. D. 1957. Hyperbolic systems of conservation laws II. *Comm. Pure Appl. Math.* 10: 537–66.

Lax, P. D. 1968. Integrals of nonlinear equations of evolution and solitary waves. *Comm. Pure Appl. Math.* 21: 467–90.

Lax, P. D. 1970. Nonlinear partial differential equations of evolution. *Proc. International Congress of Mathematicians, Nice*. 2: 831–40.

Lax, P. D. 1971. Shock waves and entropy. *Contributions to Nonlinear Functional Analysis*. Ed. E. H. Zarantonello. New York: Academic Press. Pages 603–34.

Lax, P. D. 1973. Hyperbolic systems of conservation laws and the mathematical theory of shock waves. *Regional Conference Series in Applied Mathematics* 11. Philadelphia: SIAM.

Leibovich, L. 1973. Solutions of the Riemann problem for hyperbolic systems of quasilinear equations without convexity conditions. To appear in *J. Math. Anal. and Applications*.

Leibovich, S. 1970. Weakly non-linear waves in rotating fluids. *J. Fluid Mech.* 42: 803–22.

Lesser, M. B. and R. Seebass. 1968. The structure of a weak shock wave undergoing reflexion from a wall. *J. Fluid Mech.* 31: 501–28.

Levinson, N. 1953. Certain explicit relationships between phase shift and scattering potential. *Phys. Rev.* 89: 755–57.

Lick, W. 1969. Two-variable expansions and singular perturbation problems. *SIAM J. Appl. Math.* 17: 815–25.

Lick, W. 1970. Nonlinear wave propagation in fluids. *Annual Review of Fluid Mechanics* 2. Eds. M. Van Dyke, et al. Palo Alto: Annual Reviews. Pages 113–36.

Lighthill, M. J. 1956. Viscosity in waves of finite amplitude. *Surveys in Mechanics.* Eds. G. Batchelor and R. M. Davies. Cambridge: Cambridge University Press. Pages 250–351.

Lighthill, M. J. 1960. Studies on magnetohydrodynamic waves and other anisotropic wave motions. *Phil. Trans. Roy. Soc.* A 252: 397–430.

Lighthill, M. J. 1965a. Group velocity. *J. Inst. Maths. Applics.* 1: 1–28.

Lighthill, M. J. 1965b. Contributions to the theory of waves in non-linear dispersive systems. *J. Inst. Maths. Applics.* 1: 269–306.

Lighthill, M. J. and G. B. Whitham. 1955. On kinematic waves: I. Flood movement in long waves. II. Theory of traffic flow along crowded roads. *Proc. Roy. Soc.* A 229: 281–345.

Lin, C.-C. 1955. *The Theory of Hydrodynamic Stability.* Cambridge: Cambridge University Press.

Long, R. R. 1953. Some aspects of the flow of stratified fluids. I. A. theoretical investigation. *Tellus* 5: 42–58.

Long, R. R. 1965. On the Boussinesq approximation and its role in the theory of internal waves. *Tellus* 17: 46–52.

Longuet-Higgins, M. S. 1962. Resonant interactions between two trains of gravity waves. *J. Fluid Mech.* 12: 321–32.

Longuet-Higgins, M. S. 1968. On the trapping of waves along a discontinuity of depth in a rotating ocean. *J. Fluid Mech.* 31: 417–34.

Longuet-Higgins, M. S., and O. M. Phillips. 1962. Phase velocity effects in tertiary wave interactions. *J. Fluid Mech.* 12: 333–36.

Longuet-Higgins, M. S., and N. D. Smith. 1966. An experiment on third-order resonant wave interactions. *J. Fluid Mech.* 25: 417–36.

Love, A. E. H. 1927. *A Treatise on the Mathematical Theory of Elasticity.* (4th ed.) Cambridge: Cambridge University Press. Also, New York: Dover.

Luke, J. C. 1966. A perturbation method for nonlinear dispersive wave problems. *Proc. Roy. Soc.* A 292: 403–12.

MacCamy, R. C., and V. J. Mizel. 1967. Existence and nonexistence in the large of solutions of quasilinear wave equations. *Arch. Rat. Mech. Analysis* 25: 299–320.

Martin, S., W. Simmons, and C. Wunsch. 1972. The excitation of resonant triads by single internal waves. *J. Fluid Mech.* 53: 17–44.

McGoldrick, L. F. 1965. Resonant interactions among capillary-gravity waves. *J. Fluid Mech.* 21: 305–31.

McGoldrick, L. F. 1970. An experiment on second-order capillary-gravity resonant wave interactions. *J. Fluid Mech.* 40: 251–71.

McGoldrick, L. F., O. M. Phillips, N. E. Huang, and T. H. Hodgson. 1966. Measurements of third-order resonant wave interactions. *J. Fluid Mech.* 25: 437–56.

Mei, C. C. 1966. Nonlinear gravity waves in a thin sheet of viscous fluid. *J. Math. and Phys.* 45: 266–88.

Meyer, R. E. 1967a. Note on the undular jump. *J. Fluid Mech.* 28: 209–21.

Meyer, R. E. 1967b. On the approximation of double limits by single limits and the Kaplun extension theorem. *J. Inst. Maths. Applics.* 3: 245–49.

Meyer, R. E. 1967c. Near-steady transition waves of a collisionless plasma. *J. Math. Phys.* 8: 1676–84.

Miles, J. W. 1961. On the stability of heterogeneous shear flows. *J. Fluid Mech.* 10: 496–508.

Miles, J. W. 1963. On the stability of heterogeneous shear flows, Part 2. *J. Fluid Mech.* 16: 209–27.

Miles, J. W. 1968a. Lee waves in a stratified flow: Part 1. Thin barrier. *J. Fluid Mech.* 32: 549–68.

Miles, J. W. 1968b. Lee waves in a stratified flow: Part 2. Semicircular obstacle. *J. Fluid Mech.* 33: 803–14.

Miura, R. M. 1968. Korteweg-deVries equation and generalizations. I. A remarkable explicit nonlinear transformation. *J. Math. Phys.* 9: 1202–4.

Miura, R. M., C. S. Gardner, and M. D. Kruskal. 1968. Korteweg-deVries equation and generalizations. II. Existence of conservation laws and constants of motion. *J. Math. Phys.* 9: 1204–9.

Miura, R. M., and M. D. Kruskal. 1973. Application of a nonlinear WKB method to the Korteweg-deVries equation. To appear in *SIAM J. Appl. Math.*

Morawetz, C. S., and W. Strauss. 1972. Decay and scattering of a nonlinear relativistic wave equation. *Comm. Pure Appl. Math.* 25: 1–31.

Morton, K. W. 1962. Large-amplitude compression waves in an adiabatic two-fluid model of a collision-free plasma. *J. Fluid Mech.* 14: 369–84.

Mowbray, D. E., and B. S. H. Rarity. 1967. A theoretical and experimental investigation of the phase configuration of internal waves of small amplitude in a density stratified fluid. *J. Fluid Mech.* 28: 1–16.

Nariboli, G. A. 1969. Nonlinear longitudinal dispersive waves in elastic rods. Engineering Research Institute Preprint 442, Iowa State University. (Unpublished).

Nishida, T. 1968. Global solution for an initial boundary value problem of a quasilinear hyperbolic system. *Proc. Japan Acad.* 44: 642–46.

Oleinik, O. A. 1957. Discontinuous solutions of non-linear differential equations. *Usp. Mat. Nauk* 12 (3): 3–73. English Translation: *AMS Translations* (Series 2) 26: 95–172.

Oleinik, O. A. 1957a. On the uniqueness of the generalized solution of the Cauchy problem for a non-linear system of equations occurring in mechanics. *Usp. Mat. Nauk.* 12 (6): 169–76.

Olson, L. M. 1972. Near steady oblique shock waves in a collisionless plasma. *Phys. Fluids* 15: 1070–81.

Phillips, O. M. 1967. Theoretical and experimental studies of gravity wave interactions. *Proc. Roy. Soc.,* A 299: 104–19.

Phillips, O. M. 1968. The interaction trapping of internal gravity waves. *J. Fluid Mech.* 34: 407–16.

Roffman, E. H. 1970. Localized solutions of nonlinear wave equations. *Bull. Amer. Math. Soc.* 76: 70–1.

Rozhdestvenskii, B. L. 1960. Discontinuous solutions of hyperbolic systems of quasilinear equations. *Usp. Mat. Nauk.* 15: 59–117. English translation: *Russ. Math. Surveys* 15(6): 53–111.

Sachdev, P. L., and R. Seebass. 1973. Propagation of spherical and cylindrical N-waves. *J. Fluid Mech.* 58: 197–205.

Sandri, G. 1965. A new method of expansion in mathematical physics—I. *Il Nuovo Cimento* (Series X) 36: 67–93.

Seliger, R. L. 1968. A note on the breaking of waves. *Proc. Roy. Soc.* A 303: 493–96.

Simmons, W. F. 1969. A variational method for weak resonant wave interactions. *Proc. Roy. Soc.* A 309: 551–75.

Smoller, J. A. 1969. On the solution of the Riemann problem with general step data for an extended class of hyperbolic systems. *Mich. Math. J.* 16: 201–10.

Smoller, J. A., and J. L. Johnson. 1969. Global solutions for an extended class of hyperbolic systems of conservation laws. *Arch. Rat. Mech. Analysis* 32: 169–89.

Snodgrass, F. E. et al. 1966. Propagation of ocean swells across the Pacific. *Phil. Trans. Roy. Soc.* A 259: 431–97.

Stokes, G. G. 1876. *Smith's prize examination.* Cambridge. Reprinted in *Mathematics and Physics Papers* (1905). Vol. 5. Cambridge: Cambridge University Press. Page 362.

Su, C. H., and C. S. Gardner. 1969. Korteweg-deVries equation and generalizations. III. Derivation of the Korteweg-deVries equation and Burgers equation. *J. Math. Phys.* 10: 536–39.

Taniuti, T., and C.-C. Wei. 1968. Reductive perturbation method in nonlinear wave propagation. I. *J. Phys. Soc. Japan* 24: 941–46.

Taylor, G. I. 1910. The conditions necessary for discontinuous motion in gases. *Proc. Roy. Soc.* A 84: 371–7.

Thorpe, S. A. 1966. On wave interactions in a stratified fluid. *J. Fluid Mech.* 24: 737–51.

Truesdell, C. A., and W. Noll. 1965. *The Non-linear Field Theories of Mechanics.* Vol. III/3 of the *Encyclopedia of Physics.* Ed. S. Flügge. Berlin: Springer-Verlag.

Truesdell, C. A., and R. A. Toupin. 1960. *The Classical Field Theories.* Vol. III/1 of the *Encyclopedia of Physics.* Ed. S. Flügge. Berlin: Springer-Verlag. Pages 226–790.

Ursell, F. 1953. The long-wave paradox in the theory of gravity waves. *Proc. Camb. Phil. Soc.* 49: 685–94.

Van Dyke, M. 1964. *Perturbation Methods in Fluid Mechanics.* New York: Academic Press.

Volpert, A. I. 1967. The spaces BV and quasilinear equations. *Mat. Sbornik* 73 (115): No. 2. English Translation: *Math. USSR Sbornik* 2: 225–67.

Washimi, H., and T. Taniuti. 1966. Propagation of ion-acoustic solitary waves of small amplitude. *Phys. Rev. Letters* 17: 996–98.

Wendroff, B. 1972. The Riemann problem for materials with nonconvex equation of state. Part I: Isentropic flow. Part II: General flow. *J. Math. Anal. and Applications.* 38: 454–66, 640–58.

Whitham, G. B. 1961. Group velocity and energy propagation for three-dimensional waves. *Comm. Pure Appl. Math.* 14: 675–91.

Whitham, G. B. 1965a. Non-linear dispersive waves. *Proc. Roy. Soc.* A 283: 238–61.

Whitham, G. B. 1965b. A general approach to linear and non-linear dispersive waves using a Lagrangian. *J. Fluid Mech.* 22: 273–83.

Whitham, G. B. 1967a. Non-linear dispersion of water waves. *J. Fluid Mech.* 27: 399–412.

Whitham, G. B. 1967b. Variational methods and applications to water waves. *Proc. Roy. Soc.* A 299: 6–25.

Whitham, G. B. 1970. Two-timing, variational principles, and waves. *J. Fluid Mech.* 44: 373–95.

Wijngaarden, L. van. 1968. On the equations of motion for mixtures of liquid and gas bubbles. *J. Fluid Mech.* 33: 465–74.

Yajima, N., A. Outi, and T. Taniuti. 1966. A model of the dispersive nonlinear equation, *Prog. Theo. Phys.* 35: 1142–53.

Yih, C.-S. 1955. Stability of two-dimensional parallel flows for three-dimensional disturbances. *Quart. Appl. Math.* 12: 434–35.

Yih, C.-S. 1958. On the flow of a stratified fluid. *Proceedings of Third U.S. National Congress of Applied Mechanics.* Pages 857–61.

Yih, C.-S. 1960. Exact solutions for steady two-dimensional flow of a stratified fluid. *J. Fluid Mech.* 9: 161–74.

Yih, C.-S. 1965. *Dynamics of Nonhomogeneous Fluids.* New York: Macmillan.

Yih, C.-S. 1966. Note on edge waves in a stratified fluid. *J. Fluid Mech.* 24: 765–68.

Yih, C.-S. 1967. Equations governing steady three-dimensional large-amplitude motion of a stratified fluid. *J. Fluid Mech.* 29: 539–44.

Yih, C.-S. 1969a. A class of solutions for steady stratified flows. *J. Fluid Mech.* 36: 75–86.

Yih, C.-S. 1969b. Stratified flows. *Annual Reviews of Fluid Mechanics* 1. Eds. W. R. Sears and M. Van Dyke. Palo Alto: Annual Reviews. Pages 73–110.

Yih, C.-S. 1970. Stability of and waves in stratified flows. *Proceedings of the Eighth Symposium of Naval Hydrodynamics.* August 1970. Pasadena, Calif. Pages 219–34.

Zabusky, N. J. 1963. Phenomena associated with the oscillations of a non-linear model string. The problem of Fermi, Pasta, and Ulam. *Proc. Conf. on Mathematical Models in the Physical Sciences*, University of Notre Dame, South Bend, Ind. 1962. Eds. S. Drobot and P. A. Viebrock, Englewood Cliffs, N.J.: Prentice-Hall. Pages 99–133.

Zabusky, N. J. 1967. A synergetic approach to problems of nonlinear dispersive wave propagation and interaction. *Proc. Symp. on Nonlinear Partial Differential Equations*, University of Delaware, Newark, Del. 1965. Ed. W. F. Ames. Academic Press. Pages 223–58.

Zabusky, N. J. 1968. Solitons and bound states of the time-independent Schrödinger equation. *Phys. Rev.* 168: 124–28.

Zabusky, N. J. 1969. Nonlinear lattice dynamics and energy sharing. *Proc. International Conference on Statistical Mechanics 1968. J. Phys. Soc. Japan Supplement* 26: 196–202.

Zabusky, N. J., and C. J. Galvin. 1971. Shallow water waves, the Korteweg-deVries equation and solitons. *J. Fluid Mech.* 47: 811–24.

Zabusky, N. J., and M. D. Kruskal. 1965. Interaction of "solitons" in a collisionless plasma and the recurrence of initial states. *Phys. Rev. Letters* 15: 240–43.

Zhang Tong and Yu-Fa Guo. 1965. A class of initial-value problems for systems of aerodynamic equations. *Acta Math. Sinica* 15: 386–96. English translation: *Chinese Math.* 7: 90–101.

Index

NONLINEAR WAVES

Designed by R. E. Rosenbaum.
Composed by R. & R. Clark Ltd.,
in 10 point monotype Times 327, 3 points leaded,
with display lines in Univers 689.
Printed offset by Valley Offset, Inc.
Bound by Vail-Ballou Press
in Columbia book cloth
and stamped in All Purpose foils.